CULTURE AND COSMOS
http://www.CultureAndCosmos.org

Culture and Cosmos is published twice a year, in northern spring/summer and autumn/winter, in association with the Sophia Centre for the Study of Cosmology in Culture, University of Wales Trinity Saint David.
Contributions and editorial correspondence should be addressed to:
Editors@cultureandcosmos.org

Editor: Dr. Nicholas Campion, the Editor of *Culture and Cosmos*, School of Archaeology, History and Anthropology, University of Wales Trinity Saint David, Ceredigion, Wales, SA48 7ED, UK.
E Mail **n.campion@tsd.ac.uk**

Deputy Editor: Dr. Jennifer Zahrt
Assistant Editor: Dr. Fabio Silva
Editorial Board: Dr. Silke Ackermann, Professor Anthony F. Aveni, Dr. Giuseppe Bezza, Dr. David Brown, Professor Charles Burnett, Dr. Hilary M. Carey, Dr. John Carlson, Dr Patrick Curry Professor Robert Ellwood, Dr. Germana Ernst, Dr. Ann Geneva, Professor Joscelyn Godwin, Dr. Dorian Greenbaum, Dr. Jacques Halbronn, Robert Hand, Dr Jarita Holbrook, Professor Michael Hunter, Professor Ronald Hutton, Dr Peter Kingsley, Dr. Edwin C. Krupp, Dr. J. Lee Lehman, Dr. Lester Ness, Professor P. M. Rattansi, Professor James Santucci, Robert Schmidt, Dr. Lorenzo Smerillo, Professor Richard Tarnas, Dr. Graeme Tobyn, Dr. David Ulansey, Robin Waterfield, Dr. Charles Webster, Dr. Graziella Federici Vescovini, Dr. Angela Voss, Dr. Paola Zambelli, Robert Zoller.
Technical assistance: Frances Clynes
Copy-editing: Ian Tonothy, Marcia Butchart.

Contributors Guidelines: Please see http://cultureandcosmos.org/submissions.html
Copying: Apart from fair dealing for the purposes of research or private study, or criticism or review, as permitted under the Copyright, Designs and Patents Act 1988, no part of this publication may be reproduced, stored or transmitted in any form or by means without the prior permission of the Publisher.
Front cover: Cover of *Weltall, Erde, Mensch* (1962), (see 'Research Note', pp. 111–17, this issue), source: http://www.buchfreund.de/covers/10791/26171300.jpg

Published by Culture and Cosmos, PO Box 1071, Bristol BS99 1HE, UK.
© Culture and Cosmos 2014
Printed by Lightning Source

CULTURE AND COSMOS
www.CultureAndCosmos.org

Editor Nick Campion
Vol. 17 No. 1 Spring/Summer 2013   ISSN 1368-6534

Published in Association with
The Sophia Centre for the Study of Cosmology in Culture,
University of Wales Trinity Saint David
http://www.uwtsd.ac.uk/sophia/

Editorial

Astrology and Literature

More often than not, the researcher interested in astrology will be challenged in the academy with questions of veracity: Is astrology true? Do you believe in it?[1] Yet, researchers in literary studies are rarely asked whether or not their material is true, or whether they believe in it. Fiction provides a modicum of shelter from such enquiry. When astrology is examined along the lines of and alongside narrative, the question of truth can be sidestepped and a rich array of cultural knowledge can be explored. This issue of *Culture and Cosmos* presents a collection of articles that delve into the intersections between textuality and cultural astronomy and astrology. I have deliberately chosen not to organise the articles according to temporal logic in order to resist any teleological connotations a chronological ordering may imply.

The issue opens with a robust collection of verse concerning the discovery of new astronomical bodies at the turn of the nineteenth century. Clifford J. Cunningham and Günter Oestmann present many of these works in full while demonstrating how poetry was used as an 'intellectual tool' of science, not only to memorialise the new bodies being discovered, but also to negotiate naming rights, often along political and linguistic lines.

While these nineteenth-century writers openly touted their findings, the next article discusses a potential covert encoding of astronomical

---

[1] Nicholas Campion, *Astrology and Popular Religion in the Modern West: Prophecy, Cosmology and the New Age Movement* (Farnham: Ashgate, 2012), p. 85.

observation in poetic form. Dorian Knight looks to the structure of the Eddic myth *Hávamál*, to reveal a verse description of the lunar cycle. He shows how allegory encodes astronomical information, and in turn, how this astronomical knowledge aids in unraveling the mythological content of the narrative.

In the next article Karen Smyth discusses the role of technical astronomical and astrological expressions in medieval literature by authors such as Geoffrey Chaucer and Adelard of Bath, among others. She argues that these terms become a site both for the comprehension of the temporal cosmos and the exercise of poetic experiment and the demonstration of its mastery.

Then, Kirk Little performs a literary analysis of Washington Irving's 1832 tale, *The Legend of the Arabian Astrologer*, a tale that is curiously void of technical astrological terminology. Little situates his reading in the context of early nineteenth–century astrology in England and America, and argues that we can read Irving's short story about an Egyptian astrologer as a litmus test for the status of astrological knowledge at the time—a useful model for future research into other tales.

Moving from a fictional Egypt to a real one, Guiliano Masola and Nicola Reggiani examine a curious papyrus, dated to 194 CE, that offers insight into the role astrology may have played in everyday life in ancient Eygpt. They explore the implications of this letter between two friends, in which precise and advantageous astrological advice is dispensed concerning an economic transaction.

Finally, continuing with the theme of economics and cosmology, Reinhard Mussik presents a research note about a fascinating text from former East Germany in terms of the Marxist cosmology embedded in it.

Together these articles display the myriad angles from which one can approach the intersections of literary analysis and cultural astronomy and astrology.

Jennifer Zahrt,
Deputy Editor,
Culture and Cosmos,
School of Archaeology, History and Anthropology,
University of Wales Trinity Saint David.

Classical Deities in Astronomy: The Employment of Verse to Commemorate the Discovery of the Planets Uranus, Ceres, Pallas, Juno, and Vesta

Clifford J. Cunningham and Günther Oestmann

Abstract: From 1781 to 1807 five new planetary objects transformed our basic ideas about the nature of the solar system. The coincidental discovery of these objects during a period when classical ideas permeated European thought produced a rich variety of verse in Latin, German, French, Italian and English. For the first time, they are collected here and put into context.

1. Introduction

The discovery of Uranus in 1781, the eighth planet of the solar system, was an epochal event in late eighteenth-century astronomy. Likewise, the discovery of Ceres in 1801, now categorized as a dwarf planet, was an epochal event in early nineteenth-century astronomy. The subsequent discovery of three more minor planets (or asteroids) generated huge interest and caused great controversy about their nature. These discoveries commanded attention from a wide spectrum: poets, astronomers and royalty. From the time of the Renaissance, it was realized that 'poetry is not just a pretty arrangement of words on a page but rather an intellectual tool of vast capability'.[1]

The authors would like to acknowledge Dr. Brian Marsden (Smithsonian Astrophysical Observatory), who was involved with the early development of this paper in 2010, but died before its completion. Thanks also to Dr. John Ramsey of the University of Illinois at Chicago for some of the Latin translation. Information about Elizabeth Fergusson and her poem about Uranus is courtesy of Carla Loughlin at Graeme Park in Pennsylvania.

Clifford J. Cunningham and Günther Oestmann, 'Classical Deities in Astronomy: The Employment of Verse to Commemorate the Discovery of the Planets Uranus, Ceres, Pallas, Juno, and Vesta', *Culture And Cosmos*, Vol. 17, no. 1, Spring/Summer 2013, pp. 3–29.
www.CultureAndCosmos.org

4   Classical Deities in Astronomy

In the nineteenth century, poetry and science were 'kindred thrones' which possess the 'power' to 'lift the mind above the stir of earth and win it from low-thoughted care'.[2] This important nineteenth-century aspect of science has recently been examined.[3] While poetry relating to the stars has been the subject of scholarly study, the application of poetry to memorialize the discovery of Uranus and the four asteroids has never been considered.[4] This paper addresses that neglected issue. The verses quoted here, from the late eighteenth and early nineteenth centuries, often reflect the intense rivalries between nations at the political level, and between astronomers individually as they jockeyed for the right to name these new and exciting celestial objects.

2. Eighteenth-Century Precedent
The use of Latin verse to mark advances in astronomy had eighteenth-century precedents. It was used to commemorate famous astronomers and to celebrate important discoveries. When the astronomer Tobias Mayer died in 1762, his elegy was written by Abraham Kaestner (1719–1800). Included in this elegy was a six-line Latin verse.[5]

> Te maris et terrae et magni sine limite coeli
>   Mensorem cohibent, Mayere,
> Pulveris exigui prope clausum parvula templum
>   Munera: nec quidquam tibi prodest
> Rexisse errantem lunam, movisseque summo
>   Sidera fixa polo, morituro!

---

1 Marl A. Peterson, *Galileo's Muse: Renaissance Mathematics and the Arts.* (Cambridge, MA: Harvard University Press, 2011).

2 William R. Hamilton, 'Introductory Lecture on Astronomy. Delivered in Trinity College, Dublin. November 8th 1832', *The Dublin University Review and Quarterly Magazine*, Vol. 1, (1833): pp. 72–85.

3 Gillian Jane Daw, 'The Victorian Poetic Imagination and Astronomy: Tennyson, De Quincey, Hopkins and Hardy', (PhD thesis, University of Sussex, 2011).

4 P. Boitano, 'Poetry of the Stars', *The Inspiration of Astronomical Phenomena* VI, Astronomical Society of the Pacific Conference Series, Vol. 441 (2011), pp. 289–309.

5 *The Monthly Magazine*, (Dec. 1799), p. 888.

This is based on Horace's Ode 1.28:

> Te maris et terrae numeroque carentis harenae
>   mensorem cohibent, Archyta,
> pulueris exigui prope latum parua Matinum
>   munera nec quicquam tibi prodest
> aerias temptasse domos animoque rotundum
>   percurrisse polum morituro.

For this there is an English translation by Charles Carrington:

> The sea, the earth, the innumerable sand,
>   Archytas, thou couldst measure; now, alas!
> A little dust on Matine shore has spann'd
>   That soaring spirit; vain it was to pass
> The gates of heaven, and send thy soul in quest
>   O'er air's wide realms; for thou hadst yet to die.

With similar poetic license, Kaestner's verse can be rendered as follows:

> The sea, the earth, the boundless universe,
>   Mayer, thou could'st measure; now, alas!
> The paltry gift of a little dust confines your remains next to
>   a shut-up temple; vain it was to fix the course of the wand'ring moon
> and to chart the fixed stars in the dome of the sky;
>   for thou had'st yet to die.

The line about the wandering moon is a reference to Mayer's lunar tables.[6] The line about charting the fixed stars is a reference to Mayer's catalogue of zodiacal stars.[7]

This particular Latin verse was chosen for this paper because of its dual link with the future discovery of Ceres. The author of the verse, Kaestner,

---

6 Tobias Mayer, *Novae tabulae motuum solis et lunae*. Commentarii societatis Regiae Scientiarum Gottingensis, Vol. 2 (Göttingen, 1752). Eric Gray Forbes, 'Tobias Mayer's lunar tables', *Annals of Science*, Vol. 22, no. 2, (1966): pp. 105–16.

7 Tobias Mayer, *Opera inedita* (Göttingen, 1755).

6   Classical Deities in Astronomy

was the teacher of Carl Gauss (1777–1855), who developed the mathematical method required to predict the position of Ceres at the end of 1801. This prediction, accurate to about one degree, was sufficient to enable its recovery by Baron Franz von Zach (1754–1832) and Wilhelm Olbers (1758–1840).[8]

It was also chosen because it commemorates Tobias Mayer. On the night of 1 January 1801 Giuseppe Piazzi (1746–1826) was concentrating on Francis Wollaston's star catalogue.[9] He was looking for Wollaston's star 'Mayer 87' and realized that the position given didn't agree with the 'Mayer 87' in Mayer's star catalogue. The real Mayer 87 (μ Arietis) is Piazzi II:153. By searching for Wollaston's 'Mayer 87' he found both it and Ceres![10]

3. The Discovery of Uranus

When William Herschel (1738–1822) discovered a new planet beyond Saturn in 1781, he was slow to select a name. Herschel eventually decided to name his discovery Georgium Sidus, 'The Georgian Star', in honour of his patron, King George III of Great Britain. He may have been led to this choice by Horace's phrase Julium Sidus, in compliment to Julius Caesar.[11]

Since no one outside of Great Britain accepted the appellation, many other names were proposed: Herschel (favoured by the French astronomer Pierre-Simon Laplace), Astraea, Cybele, Neptune, Hypercronius, Transaturnis and Minerva, among others. The multiple names attached to the recent solar system discoveries (Uranus=The Georgian Planet=

---

8 Clifford J. Cunningham, *The First Asteroid* (Ft. Lauderdale, FL: Star Lab Press, 2001). Peter Brosche, 'Die Wiederauffindung der Ceres 1801', in *Astronomie von Olbers bis Schwarzschild*, eds. Wolfgang R. Dick and Jürgen Hamel. Acta Historica Astronomiae Vol. 14. (Frankfurt am Main: Deutsch, 2002), pp. 80–88.

9 Francis Wollaston, *Specimen of a General Astronomical Catalogue arranged in zones of north polar distance* (1789).

10 Clifford J. Cunningham, Brian G. Marsden, and Wayne Orchiston, 'Giuseppe Piazzi: The Controversial Discovery and Loss of Ceres in 1801', *Journal for the History of Astronomy,* Vol. 42, no. 3 (2011a): pp. 283–306.

11 Letter from Herschel to Sir Joseph Banks, 7 November 1782. In the Herschel Archives, Royal Astronomical Society, London. Horace, Lib. I, Ode XII, lines 45-48.

Herschel; Ceres=Hera=Piazzi) also found its parallel in botany, where the same species might have several designations.[12]

The French classical scholar Louis Poinsinet de Sivry (1733–1804) chose Cybele, the spouse of Saturn and daughter of Uranus. He represented the order of the planets in the following verse.[13]

> Ambit Solem Hermes, Venus hunc, mox Terra, Diana.
> Mars sequitur. Pergit Rex Juppiter. Hunc Saturnus.
> Omnes hos orbes amplectitur alma Cybele.

> Mercury orbits the Sun; Venus orbits beyond Mercury, and next the Earth, the Moon.
> Mars follows. King Jupiter makes his way next. And Saturn orbits beyond Jupiter.
> All these orbs are enclosed by nurturing Cybele.

The appellation 'Georgian Star' was the only one accepted in England, and it was celebrated in poetry and an epigram. In this extract from a long poem about the year 1782, the 'thou' referred to is the Muse of Glory.[14]

> Thou, who each Planet in his Orbit guide'st,
> While round the Sun, on wings of light, thou ride'st,
> Stop, ruling Angel, in thy rapid round,
> And, at thy Solar-System's utmost bound,
> For one short moment, from thy native skies,
> View the concluding Year with fav'ring eyes:
> Beyond the search of NEWTON's heav'nly eye,
> Behold ambitious HERSCHEL dare to spy
> (Aided by wond'rous Optic Glass) from far
> The dim faint splendours of the GEORGIAN STAR.

---

12 J. S. Phillips, 'Nomenclature of Natural Science', *Proceedings of the Academy of Natural Sciences of Philadelphia*, Vol. 1, (1841), pp. 85–88.

13 *Monatliche Correspondenz*, (July 1801), p. 66.

14 William Tasker, *Annus mirabilis; or, The eventful year eighty-two, An historical poem*. Verse 819–827. (Exeter: B. Thorn and Son, 1783).

8   Classical Deities in Astronomy

The name was also reiterated in this brief epigram.[15]

>GEORGIUM SIDUS, the new-discovered Planet
>
>BRITAIN, in spite of ev'ry blow,
>   Thy George superior still shall rise;
>Fate lessen'd here his realms below,
>   And gave him kingdoms in the skies.

The unseemly dispute over the naming of Herschel's discovery inspired Georg Szerdahely (1740–1808) to compose a verse in 1787. Szerdahely was a professor of rhetoric who taught aesthetics at the University of Buda (modern-day Budapest). The verse was reworked by Baron Franz von Zach (editor of the *Monatliche Correspondenz*) in 1802 to describe the Ceres nomenclature controversy.[16]

>The astronomers battle, and still the case is before the judge,
>with which name a new planet should be designated.
>   *He has to exclaim once again:*
>
>O Gods above! What would this confusion of voices be?
>If each voice should offer a name!
>   *We shall wait and see and cry:*
>
>It is not ours to settle the disputes of voices that have already begun.

---

15 *The London Magazine*, Vol. 3 (1784), p. 297.

16 Lines in italics added in 1802 by Zach. Clifford J. Cunningham, Brian G. Marsden, and Wayne Orchiston, 'How the First Dwarf Planet Became the Asteroid Ceres', poster paper presented at International Astronomical Union conference in Rio de Janiero, 2009.

Another German publication devoted a full page to an anonymous poem about the discovery.[17]

Der neue Planet

Der Deutsche, der, an Avons Strand,
Des Himmels jüngsten Liebling fand,
Grüßt' ihn entzückt: 'Georgia!'
Damit in dieser weiten Sphäre
Dem besten Herrn Amerika,
Das ihn verließ, ersezet wäre.

Pfui, rief der Franze, nimmermehr
Soll unsres alten Feindes Ehr'
In Ewigkeiten prangen! Da,
Bei des Gemals bekanten Grenzen,
Wird eher Cybeleia
Im neuen Irgestirne glänzen.

Warum willst du den heil'gen Ort,
So fiel ihm Bode rasch ins Wort,
Nach Deines Volks Galanterie'n,
Ihr weihen? Dort in niedren Höhen
Siehst du sein Harem sich um ihn
Im engren Kreise leuchtend drehen.

Am besten machet Uranus
Vom Fabelstamme den Beschluß.
In seine Reihe schicket sich
Kein Fürst vom sterblichen Geschlechte;
Wer hätte sonst als Friederich
Zur ersten Stelle größere Rechte?

---

17 *Deutsches Museum,* Vol. 11, (Leipzig, June, 1786), p. 563.

The New Planet

The German, who, on Avon's beach,
Found sky's youngest darling,
Greets him delighted: 'Georgia!'
So that in this vast sphere
Worthy Mr. America,
Who abandoned him, should be replaced.

Fie, the French exclaimed, nevermore
Should our old foe's honour be
Emblazoned in eternity! There,
Near the bridegroom's known borders,
Rather Cybele may
Glare in the new planet.

Why will thou the holy place
So Bode cut off rashly,
Hallow after your people's gallantries?
There in lower heights
You see his harem around him
In close circles revolving brightly.

Uranus of fabulous root
Makes best the close.
In his succession
No ruler of mortal lineage shall acquiesce.
Who else than Frederick
Should have greater rights in first place?

The first verse refers to Herschel as a German (he was born in Hannover) on the shores of Avon, a reference to England. This may also be a reference to the river Avon, which runs through the city of Bath, from where Herschel discovered Uranus. The first verse also makes a pointed reference to the fact that England's King George III (1723–1820) was abandoned by America. The second verse gives voice to the French opinion about naming the planet after George, giving instead their choice of Cybele. The third verse refers to Johann Bode (1747–1826), Director of Berlin Observatory, who proposed the name Uranus. The writer concludes by saying that a classical deity is a much better choice than one of mortal origins, slyly noting that none other than Frederick the Great (1712–1786) of Prussia should be given priority over any other mortal (including King George III) if a planet should be named for one.

The discovery of Uranus also elicited great interest in the newly-formed United States of America, where the Revolutionary War was still raging. In the small town of Horsham, Pennsylvania, Elizabeth Fergusson *née* Graeme (1737–1801) hosted literary 'Attic Evenings' in Keith House. The house, located in Graeme Park, is the only surviving residence of a Colonial Pennsylvania Governor. Elizabeth, who published poetry in Philadelphia magazines and newspapers, was inspired to pen a poem about Uranus entitled 'Upon the Discovery of a Planet by Mr. Herschel of Bath and by Him Named the Georgium Sidus in Honor of His Britannic Majesty'.

It appears in her Commonplace Book (somewhat like a diary, but usually written to be read by others), compiled in 1787 for Annis Boudinet Stockton. The antipathy felt by the author towards King George III, who was then trying to recapture America for the British Crown, is pointedly expressed in her verse.[18]

> Whether the optics piercing eyes
> Have introduc'd to view,
> A distant planet of the skies,
> Bright, wonderful, and new?
>
> Or whether we are nearer thrown
> To the grand fount of light,
> And from that source each mist is flown
> That wrapt the star in night?
>
> To deep this point, a female pen,
> Dare not such heights explore;
> The subject's left to learned men,
> Of philosophic lore!

---

18 The book is a compilation of things she wrote between 1770 and 1787. The note at the end says that it was written at Graeme Park on 6 January 1784 and is signed 'Laura' (a pen name she frequently used). She also wrote at the end of the entry 'This was printed in the Newspaper but not with the signature Laura'.

A star is found, that's clear; and hail'd
With Britain's monarch's name;
If his terrestrial glory's fail'd,
The Heaven's enroll his name.

But sordid Souls I greatly fear,
Will not the Change approve:
To think the Empire fled from here,
In Azure plains to rove.

Perchance in Days to Come some youth
Whose Bosom genius fire
When warmed with Scientific Truth,
He ardent thus Enquires.

What mortal Great who dwelt on Earth
Assigned this Star his Name?
Another George of Martial Worth
May be mistook by Fame.

Yet be it fixed Britannia's King,
We with this Planet done;
Will yield the late found Star to him
And Hail our George a Sun.

In France, the use of verse replete with classical allusions to celebrate scientific advancements was deeply entrenched.[19] Of note are the astronomical poems of Dominic Ricard (author of *La Sphere* in 1796; died 1803) and Louis Marcelin de Fontanes (author of *Essai sur l'Astronomie* in 1789; 1757–1821). Herschel's discovery was commemorated in French by Gudin de la Brenellerie (1738–1820), associate member of the Institut National.[20]

---

19 Jean Dhombres, 'Culture scientifique et poésie aux alentours de la Révolution française', in *Nature, histoire, société : Essais en hommage à Jacques Roger.* Rassemblés et présentés par Claude Blanckaert, Jean-Louis Fischer, Roselyne Rey, (Paris: Klincksiek, 1995).

20 Paul-Philippe Gudin de la Brenellerie, 'L'Astronomie,' poème en quatre chants (2nd edition; first was published as 'poème en trois chants' in 1801). (Auxerre, L. Fournier, an IX., 1810).

L'amour propre si vif, et si souvent déçu,
Pretendait dans les cieux avoir tout apercu;
Quand soudain on apprend du fond de l'Angleterre,
Qu'il s'offre un nouvel aster aux regards de la terre;
Que par de-là Saturne il brille dans la nuit;
Qu'Herschel l'a découvert, qu'il l'observe et le suit.

Self-esteem so lively, and so often disappointed,
Pretended to have seen everything in the skies;
When all of a sudden one learned from the heart of England,
That a new star presented itself to the world;
That beyond Saturn it shines in the night;
That Herschel had discovered it, observes and follows it.

The British were quite proud that a planet had been named for their monarch, as evidenced by this verse commemorating the discovery of the moons of Uranus.[21]

Delighted Herschel, with reflected light,
Pursues his radiant journey through the night;
Detects new guards, that roll their orbs afar,
In lucid ringlets round the Georgian star.

The 'delighted Herschel' was also personally involved in the inclusion of astronomical discoveries in verse. He worked closely with Dr. Charles Burney in the creation of a lengthy astronomical poem, which Burney later abandoned and largely destroyed.[22] It has also been noted recently that certain astronomical imagery in the grand verse-drama *Prometheus Unbound* by Percy Bysshe Shelley most likely was inspired by Herschel's cosmological discoveries.[23]

The highly charged political nature of Herschel's choice 'Georgian' resonated through the decades of the conflict with France. After caustically

---

21 Erasmus Darwin, *The Temple of Nature* (Baltimore: John Butler, 1804).

22 Roger Lonsdale, *Dr. Charles Burney: A Literary Biography* (Oxford: Clarendon Press, 1965).

23 Christopher Goulding, 'Shelley's Cosmological Sublime: William Herschel, James Lind and "The Multitudinous Orb"', *The Review of English Studies*, New Series, Vol. 57, no. 232 (2006): pp. 783–92.

noting that French astronomers had renamed the belt and sword of Orion in honour of Napoleon Bonaparte (1769–1821), the writer (named only as Stella), concluded the poem thus:

> The rays of *Orion* oft guide our bold tars;
> But they ne'er will be led by *'Napoleon's Stars.'*
> We discover'd a planet, and call'd it *our own*,
> As a tribute to virtue that beams on the throne;
> But they, who the Georgium Sidus deride,
> New name the *old* stars to please *Corsican* pride.[24]

4. Anticipating a New Planet

In his 1773 didactic poem 'Sistema dei Cieli' (System of the Skies), Carlo Gastone della Torre Rezzonico of Como (1742–1796) wrote about a little unknown planet between Mars and Jupiter.[25] He was quite correct in surmising that it was both the smallness of its disk, and its low albedo, that had prevented it from being seen. The largest asteroid Ceres has a diameter of 960 km and a geometric albedo of 0.09. This compares to the smallest planet Mercury, with a diameter of 4878 km and an albedo of 0.138, lowest among the major planets.

> Sola poi vien la rubiconda stella
> Del Fero Marte e dopo lui l'immenso
> Giove, che tanto gli è lontan quant'esso
> Dal Sol due volte. In così vasto campo
> Forse alcun'altra dell'erranti stelle
> Ruota da noi non conosciuta, e forse
> Suo picciol disco, o per gran macchia oscuro
> Fe sì, che invan della ritrosa in cerca
> Al notturno favor di doppia lente
> Vagò pel ciel l'astronoma pupilla...

---

24 On 28 August 1807 this poem entitled 'Napoleon's Stars' was published in England in the *Anti-Jacobin Review*, Vol. 27, p. 528.

25 Piero Sicoli, 'Duecento anni fa la scoperta di Cerere', *l'Astronomia*, no. 214, (November, 2000), p. 30. Milan.

Alone then arrives the reddish star of the proud Mars,
and after it the huge Jupiter, that is twice as far from it as
it is from the Sun. In so vast space maybe some
other of the wandering stars revolve unknown to us and
maybe because of its little disk, or because of darkness stain,
made so that in vain was a search of the bashful made
with the favoured nocturnal double lens
as the astronomer's pupil roamed the sky....

5. The Discovery of Ceres

From the Palermo Observatory in Sicily, Giuseppe Piazzi discovered an extraordinary object on the first day of the nineteenth century, and four months later he had christened it Ceres Ferdinandea. Ceres was chosen as the patron goddess of Sicily, and Ferdinand in honour of Piazzi's patron King Ferdinand of Naples and Sicily.[26] Once it had been recognized by the astronomical community as a new planet, King Ferdinand felt obliged to commemorate the event. He first proposed to strike a gold medal, but was dissuaded in this intention by Piazzi, who asked that the funds instead be used for astronomical instruments.[27]

In England, it was dryly noted that 'the King of Naples has added sixty pounds a year to Mr. Piazzi's salary, for the discovery of the new planet, and honouring it with the royal name. So small a reward assuredly justifies astronomers in refusing to accede to the new title, and in immortalising the discoverer rather than the monarch'.[28]

Since the commemoration of the discovery was thus somewhat subdued in Sicily, it was left to those with the power of verse to mark the event for the ages. One who took the opportunity to do so was the Italian poetic improviser Pietro Scotes from Verona, who was quite the sensation just after the turn of the century in Weimar and elsewhere in Germany.

'The various themes he [Scotes] set for himself to render in various poetic meter, in ottava rime etc., included: the advantages of blondes over brunettes, Achilles' lament for Patroclus, Nina's lament for her beloved,

---

26 Clifford J. Cunningham, Brian G. Marsden, and Wayne Orchiston, 'How the First Dwarf Planet Became the Asteroid Ceres', *Journal of Astronomical History and Heritage*, Vol. 12, no. 3, (2009): pp. 240–48.

27 Cunningham, *The First Asteroid*, p. 193.

28 *Annual Review of History and Literature* for 1804, Vol. 3 (1805), p. 855.

the advantages of music over painting, and of hope over fulfillment. One of his most beautiful poems was dedicated to the discovery of Ceres Ferdinandea, whereby he took every opportunity to extol the merits of his fatherland'.[29]

Unfortunately it appears the text of his poem on Ceres has not survived. 'All these things were extemporized at (often exclusive) social gatherings (what professional musicians today call "one nighters"), with individual poems often prompted by a topic, line, meter, or even end rhyme supplied by the audience or guest of honor—but the poems themselves were to my knowledge neither written down nor published. The reviews (there are two) address his performance rather than the text of the extemporized poems'.[30]

5.1 Verses in the *Monatliche Correspondenz*

As Editor of the world's only astronomical journal, the *Monatliche Correspondenz* (Monthly Correspondence), Baron Franz von Zach was in a unique position to publish whatever he saw fit. Not content with printing positional measurements of Ceres, he often inserted personal comments and quoted directly from the letters he received.[31] 'One of my friends expresses the order of the now eight planets in the following not unsuccessful verses, which, according to the custom of usual memorial verses, expresses a further thought'. Here the name Hera (spouse of Zeus in the Greek pantheon) is used instead of Ceres. Anticipating the discovery of a new planet, it was the name selected by Zach's patron, Duke Ernst II of Saxe-Gotha, 16 years earlier.[32]

> Mercurius primus; Venus altera; Terra deinde;
> Mars posthac; quintam sedem sibi vindicate Hera.
> Juppiter hanc ultra est. Sequitur Saturnus; at illum
> Uranus egreditur, non ausim dicere summus.

---

29 Anon., Review of 'Der Improvisator Scotes', *Der neue Teutsche Merkur*, Vol. 3, no. 9, (September 1802): pp. 71–73. Karl August Boettiger, 'Der Improvisator Pietro Scotes aus Verona', *Der neue Teutsche Merkur*, Vol. 3, (1802): pp. 135–48.

30 Doug Stott, Personal communication (2011).

31 Clifford J. Cunningham, et al., 'Giuseppe Piazzi' (2011), pp. 283–306.

32 Anonymous verse, *Monatliche Correspondenz*, (July 1801), p. 67.

Oder:
Mercurius Solem comitatur proximus. Illum
Insequitur Venus, hano Tellus, Luna comitante;
Mars posthac, Martem prohibit Jovis esse sequacem
Hera lateens srustra, et melioribus obvia vitris.
Saturnum extrema Proavi statione locabant,
Nos aliter. Supremara coeli nunc Uranus arcem
Usurpat, poenas ausi fortasse daturus.

Mercury first, Venus second, then the Earth;
Mars after Earth; Hera lays claim to the 5th place.
Jupiter is beyond that one. Saturn comes next; but
Uranus (I should scarcely dare to say the last) makes his way beyond Saturn.

Or:
Mercury is the closest companion of the Sun.
Venus follows Mercury, and Venus is followed by Earth, with its companion the Moon;
Mars comes after Earth; Jove forbids Mars to be a close follower.
Next is Hera hiding in vain and exposed by better lenses.
Earlier generations situated Saturn in the outermost place,
But not we. Uranus now lays claim to the farthest arc of heaven,
Destined, perhaps, to be punished for his daring deed.

It is interesting to note that this is the only one of the verses dedicated to the discovery of Ceres that mentions—albeit rather obliquely—the technology that made it possible (i.e., 'better lenses'). Likewise, the Ramsden Circle used by Piazzi to discover Ceres did not feature in any paintings or engravings done to commemorate the event, although it was depicted in relation to Piazzi's star catalogue.[33]

Zach rhapsodized about the discovery of both Ceres and Pallas, inserting what appears to be a Latin verse of his own design. 'It is easy to

---

33 Clifford J. Cunningham, Brian G. Marsden, and Wayne Orchiston, 'The Attribution of Classical Deities in the Iconography of Giuseppe Piazzi', *Journal of Astronomical History and Heritage*, Vol. 14, no. 2, (2011b): pp. 129–35.

18   Classical Deities in Astronomy

immortalize such an epoch-making occurrence in the history of astronomy. The heavens will proclaim these works to all people and for all time to come'.[34]

> Videbo coelos tuos, opera digitorum tuorum,
> Cerem et Palladem, quae tu fundasii.
>
> I will see thy heavens, the works of thy fingers,
> Ceres and Pallas, which thou hast founded.

5.2 Capel Lofft
The English antiquarian Capel Lofft gave public vent to his long-held fascination with astronomy in several sonnets.[35] His lengthy poem 'Eudosia' encompassed most of what was known in astronomy on the eve of the discovery of Uranus.[36] On 29 August 1801 he penned a sonnet about the newly-discovered planet:[37]

---

34 *Monatliche Correspondenz*, (June 1802), p. 589.

35 Roger Meyenberg, *Capel Lofft and the English Sonnet Tradition* (Tübingen: Francke, 2005).

36 Capel Lofft, *Eudosia: or, A Poem on the Universe*. (London: W. Richardson, 1781).

37 Capel Lofft, *Laura, or, an Anthology of Sonnets*. (London: R. & A. Taylor, 1814). This includes the exact date of creation and a dedication to Miss Finch, who became Lofft's wife. The verse, without her name, was originally published in the *Monthly Mirror* (December 1801), p. 416.

To Miss Sarah Watson Finch.

With a Sketch of THE SOLAR SYSTEM, according to the latest Discoveries.

On the supposition of a new-discover'd Planet..

To thee whom as MINERVA* I revere,
To whom my cares and happier thoughts all tend,
This sketch of every planetary Sphere
Known to obey our central Sun I send.

In these the eccentric orbs have ear
To Harmony divine! the wild career
Of Comets thus revolves: prompt to descend
To that great source which rules their mighty year.

O might my Griefs and my charm'd Passions hear
Like influence divine! - thus should I know
Like thee to teach my moments how to flow
Useful and calm; unrackt by Doubt and Fear,
And thus ascend above all earthly Woe;
That Order, Heaven's bright Grace, anticipating here.
                                              C.L. 1801.

*It was hop'd the New Planet, if ascertain'd to be such, would be nam'd Minerva: in conformity to the other mythologic designations, and in honour of Science, and of the Arts of Peace.

5.3 Nicolaus Lipari
In 1801 Piazzi sent a copy of his treatise about Ceres to Lofft, together with a Latin epigram by a Sicilian, Nicolaus Lipari. Here is how Lofft described it:[38]

---

38 Giuseppe Piazzi, *Risultati delle Osservazioni della Nuova Stella*. (Palermo: Palermo Observatory, 1801). Lofft, Capel, *Monthly Mirror* (April, 1802), p. 240–41.

"If these Observations will be acceptable for the Mirror they are much at your service, as also the subjoin'd Epigram on its discovery and name, CERES FERDINANDEA, which I think has not appear'd in Print in England. Piazzi has prefixt it to his Account."

Alma Ceres, pertaesa hominum consortia, summas
Ut Superum tetigit, non reditura, Domos,
Septem inter Caeli volventia sidera, cursum
Flectere, et immensas caepit inire vias;
Mortales fugiens oculis! Post saecula tandem
Longa, ubi conspectum non renuisse datum est,
Ante alias SICULAE voluit nova fulgere Terrae
Immemor haud Patriae, quae sibi culta, suae.
                                        Nicolaus Lipari.

Ceres from Human intercourse had fled
And viewless through the Heavens her orb had led
Mid seven companion Planets fond to stray
Latent, through the immense aerial way,
When, after Ages, to our sight was given
This last-discover'd Daughter of our Heaven.
As chief Sicilia, while on earth, she blest,
On SICILY her STAR first shone confest.
                                        C.L.

5.4 Marcin Odlanicki Poczobut

The 73-year-old Polish astronomer Marcin Poczobut (1728–1810) was an assiduous observer of Ceres from Vilnius Observatory in Lithuania. His colleague at Cracow Observatory, Jan Sniadecki, kept Zach apprised of Poczobut's work. A letter from Sniadecki to Zach dated 24 May 1802 includes more than just positional data from Poczobut. 'He loves to write Latin verses and sometimes quite good ones. You will find at the beginning of his observations two Latin verses about the distinctive character of Ceres'.[39]

> Quae segetum culmos docuisti falce secare
> Falx dentate sacrum sit tibi stemma Ceres.

> Thou hast taught her to cut the stalks
> Of standing corn with a sickle.
> The toothed sickle shall become for you
> The consecrated garland of Ceres.

This verse refers to the use of the sickle (suggested by Zach) as the planetary symbol to denote Ceres.[40] This matter is discussed in more detail in Section 6.

5.5 Michel Monti
The Piarist monk Michel Angelo Monti (1751–1822) gave the reason for the naming of Ceres in Latin verse.

> Telluris patraie ductura a Principe nomen
> Astra inter Siculis fulsit ab axe Ceres.[41]

> From the most important of the fatherland of the Earth
> the name will be derived
> Immortality shone from the eye of Ceres
> among the Sicilians.

Monti was a poet and orator. A native of Genoa, he became Professor of Eloquence in the University of Palermo. Thus his line about Sicily being 'the most important fatherland of the Earth' is an homage to his place of

---

39 *Monatliche Correspondenz*, (July 1802), p. 63. See also Clifford J. Cunningham, *The Collected Correspondence of Baron Franz von Zach*, Vol. 1, (Ft. Lauderdale, FL: Star Lab Press, 2004).

40 B. Boncompagni, 'Intorno ad una lettera di Carlo Federico Gauss al Dr. Guglielmo Mattia Olbers Memoria', *Atti dell'Accademia Pontificia de' Nuovi Lincei*. Tomo XXXVI (Rome, 1882–1883), pp. 2012–95. Citation from p. 265.

41 *Monatliche Correspondenz*, (Jan. 1811), p. 7.

residence. There is a monument to him in the San Domenico church in Palermo, the city in which he died at age 71.

6. The Discovery of Pallas

On 28 March 1802, Wilhelm Olbers discovered another 'planet' between Mars and Jupiter. Named Pallas in honour of the goddess Pallas Athena, it is now designated as 2 Pallas, signifying it as the second minor planet or asteroid. Unlike Ceres, Pallas has not been accorded 'dwarf planet' status in the twenty-first century, as it is too small to qualify.

The discovery of Ceres, the first object found between Mars and Jupiter, naturally elicited a substantial outpouring of commemoration in verse. Pallas received very little notice by comparison. In addition to his verse about Ceres, Poczobut wrote about Pallas.[42]

> Falx Cereris signum esto; tu ut taere laboris
> Sideribus sacros, aegida Pallas habe.
>
> Oh sickle, be the sign of Ceres; and have with you
> Pallas who has the Aegida, so that you can protect
> the sacred works in the skies.

Each of the major planets had long been assigned a symbol, often used as shorthand in planetary tables. When Piazzi's discovery Ceres was named, it was suggested by Zach that a sickle be used as its sign, in accordance with the role of Ceres as goddess of Agriculture. The Aegida is the protective shield made of goat skin and the Medusa head, used by Pallas Athena.

Even though Pallas had been discovered in Bremen, the following verse by an anonymous author is lacking in any nationalistic sentiment. This could possibly be due to the fact that Germany was not yet a unified country.[43]

---

42 *Monatliche Correspondenz*, (July 1802), p. 74.

43 Berlin newspaper *Vossische Zeitung*, (16 September 1802).

Der neue Planet Pallas

Endlich erschienest du Pallas, und mit dir der Oelzweig
 des Friedens;
Göttin der Weisheit, warum nahtest du leider so spät?
Mögest du heller noch leuchten, daß heller auf Erden es
 werde!—
Siehe wie funkelt so hell Venus vor allen hervor!
Zünd' o Weisheit dein Licht am Schwesternaltare der
 Liebe!
Weisheit mit Liebe vereint—so nur beglückt sie die
 Welt.

The new planet Pallas

At last you appeared, Pallas, and with you the olive branch of
 peace;
Goddess of wisdom, why did you come so late?
May your light shine even brighter so that it might be lighter on
 Earth!—
See how bright Venus sparkles!
Light, o wisdom, your fire at the sisterly altar of love!
Wisdom combined with love—it only gladdens the world this way.

A footnote told the reader that Dr. Olbers discovered Pallas on 28 March, one day after the Amiens peace treaty (which marked the end of the French Revolutionary War), which explains the peace reference in the second line. The writer then implores Pallas to join with Venus, the goddess of love, so that their combined radiance (the wisdom of Pallas and the love of Venus) may salve the wounds of a lengthy war.

7. Juno and Vesta

When Gudin de la Brenellerie published his Astronomical Poem in 1801, only seven planets were known, but the discovery of four new 'planets' from 1801 to 1807 prompted him to revise the poem. Juno, the third one, was discovered by Karl Harding in 1804 and named in honour of Jupiter's spouse. The fourth was found by Olbers in 1807 and named Vesta, Roman goddess of the home and family. As the review in the *Mercure de France*

24 Classical Deities in Astronomy

suggested, such a profusion of new planetary objects made it quite likely that the roster of objects in the solar system was far from complete: 'Mais quatre nouvelles planètes découvertes en peu d'années, le font douter que le nombre en soit encore complet'.[44]

Gudin wrote about the unexpected developments with a sequence of questions:

> Mais ai-je tout compté? Mais puis-je être assure
> Qu'un meilleur telescope, un oeil mieux éclairé,
> Sondant des vastes cieux les profondes retraites,
> N'apercevra jamais que ces onze planets?
> Ce nombre est-il prescript? Ai-je atteint et pu voir
> Le terme où du soleil s'arrête le pouvoir?
>
> But have I counted all? But can I be sure
> that a better telescope, a better educated eye,
> scanning the vast skies[,] the deepest retreats,
> will never see but these eleven planets?
> Is this number fixed? Have I reached and been able to see
> the limit where the sun loses its power?

A lengthy Latin verse by 'Pastor Schulze zu Polenz bey Leipzig' was published in 1810. In rich detail, it encompasses the discoveries of Uranus and the four asteroids.[45]

> In media *Sol* sede regit lucemque ministrat
> Orbibus undenis cunctisque sequacibus horum.
> Illorum hos *Comites* dicunt illosque *Planetas*.
> Solem quisque sua circum pro lege rotatur
> Ocyor igne Jovis, non unguem a calle recedens.
> Orbita cuique sua est distans a Rege statuta
> Lege, minor propior majorque remotior ambit.
> *Mercurius* Regem primus circuire jubetur,
> Perque dies octo undecies sibi conficit orbem.
> Hunc ultra *Venus* est splendore et lumine praestans:
> Hebdomadas tringinta duas absolvit eundo.

---

44 Pierre Louis Ginguené, Review of the new edition of the poem L'Astronomie by Gudin. *Mercure de France,* Vol. 49, (1811), pp. 209–15.

45 *Monatliche Correspondenz,* (Dec. 1810), pp. 576–78.

Tellus cum *Luna* sequitur, data mansio nobis;
Mensibns haec bis sex praescriptum conficit orbem.
Scanditur ad *Martem*, qui lusem miscet et ignem:
Bis septem hebdomadas septenas pervolat orbem.
Huncce Jovemque inter veteres vacuum esse putarunt.
At cum nostra aetas nimia intervalla videret,
Atque a consueta distandi lege recedei:
Res suspecta viros investigare monebat;
Ingenio, arte, armis instructi, vera videbant,
Proque uno quatuor cernunt, mirabile dictu,
Fragmina quos *Olbers* rupti putat esse planetae.
Sic *Cererem* Siculus detexit forte *Piazzi*;
Hebdomadas bis sex vicenas pervolat orbem.
Pallada sic *Olbers* Bremensis acutus adivit;
Haec per idem *Cereri* tempus circumvolat orbem.
Sic se aperit Goettingensi *Harding* candida *Juno*;
Per decies quinos et tres huic est via menses.
Quaesitam felix *Vestam* sic conspicit *Olbers*;
Per quadraginta tres huic est semita menses.
Ambitus his quatuor, quo non perplexior ullus,
Vix foret explicitus, nisi *Praestantissime* nobis
Montstrasses *Gauss*, nunquam moriture. Planetas
Jupiter hos ultra es, cunctorum maxime, coeli
Tu decus excellens, dignus tu, quem comitentur
Bis bini comites, *Veneris* tu lumine fulges;
Bis fere sex annis stadium métier coruscus.
Saturnus sequitur, comites quem tres quatuorque
Circumeunt. Duplex cingit, mirabile visu,
Annulus hunc; fere ter denis iter exiget annis.
Uranus insequitur, quem tu, pater Astronomorum,
Herschel, digne, tuo qui tangas vertice soles.
Errantem agnosti primus. Distare jubetur
Ultimus ante omnes, poenas fortasse daturus;
Namque Gigantes Titanasque deumque hominumque
Terrorem genuit. Comites circum volitantes
Ter duo sunt illi, plures fortasse patescent;
Finit iter decies octonis amplius annis.

In the midst of his throne the Sun rules and furnishes light
To all the eleven spheres following.
Of these orbs, some they call Comets, and some Planets.
Each one is whirled around the sun according to its own principle
More swiftly than the fire of Jove, not drawing back its hand from the heat.
The orbit to each is established by its own principle at a distance from the King,
The lesser circle closer, and the greater more distant.
Mercury is the first commanded to go around the King,
And through eight times eleven days it finishes its circuit.
Beyond him Venus stands out with its brilliance and light;
In thirty-two weeks she completes her journeying.
The Earth, the dwelling given to us, follows along with the Moon;
She consumes her ordained orbit in twice six months.
This gives rise to Mars, who mingles light with fire;
He flies across his orbit in twice seventy-seven weeks.
Among the ancients, they judged Jupiter to be solitary.
But when our generation should look upon excessive spaces,
And be withdrawn from the usual law of standing apart:
A suspect matter advises men to search deeper;
Prepared with talent, skill, equipment, they see the truth,
They see four before the one, amazing to say,
Which Olbers judges to be fragments of a shattered planet.
Thus the Sicilian Piazzi uncovered Ceres by chance;
She flies across her circuit in twice six times twenty weeks.
Thus did wise Olbers of Bremen add Pallas;
She wings about her orbit in the same time as Ceres.
Thus did lucid Juno reveal herself to Harding at Goettingen;
Her path is three and ten times five months.
Happily did Olbers catch sight of the looked-for Vesta;
Through forty-three months she takes her road.
The circuit of these four, than which nothing is more puzzling,
Would scarcely have been explained, unless you, Gauss,
Standing so much above us, would have demonstrated it, may you never pass on.
Beyond these Planets you, Jupiter, greatest of all,
Distinguished glory of the heavens, whom your four companions accompany,
Worthily you shine with the light of Venus;

> In nearly twice six years you traverse your course, twinkling.
> Saturn follows, about whom go three and four companions.
> A two-fold ring, wondrous to look upon,
> Girdles him; He finishes his route in nearly ten times three years.
> Uranus comes along after, whom you, father of Astronomers,
> Worthy Herschel, you who touch the stars with your head,
> First recognized wandering around. He is bidden to be
> Most distant of all; perhaps it is a punishment;
> For the Giants and Titans gave birth to the terror
> Of Gods and Men. He and his six companions
> Fly about, perhaps more will be revealed;
> He finishes his path in more than ten times eight years.

The 'companions' are satellites—the four Galilean satellites of Jupiter, the four satellites of Saturn, and six of Uranus are mentioned. This latter is, however, incorrect. Herschel discovered two moons of Uranus in 1787. He then claimed two more moons in 1790 and an additional two in 1794. These latter four moons were spurious, but this was not known until the 1850s. The issue of the 'suspect matter' is covered in section 4 of this paper.

This poem is unique in mentioning, as 'fragments of a shattered planet', the asteroid explosion hypothesis of Olbers. His idea, now discredited, was that the four asteroids were originally part of a much larger planet that exploded eons ago. The hypothesis—widely accepted in the early nineteenth century—was the subject of lively debate and controversy for more than a century.[46]

The allusion to Vesta being 'looked-for' reflects the fact that Olbers did a deliberate search of the sky for more asteroids. After several years of effort, he finally found Vesta.

Better than any other, this poetic tribute encapsulated the discovery of—and mystery surrounding—the four new planets.[47]

---

46 Olbers first broached the idea of a planet shattered into fragments in a letter to Carl Gauss (23 April 1802). Bremen, State and University Library; reproduced in Carl David Schilling, ed., *Wilhelm Olbers: Sein Leben und seine Werke* (Berlin: Julius Springer, 1894–1909). How the hypothesis was received during the past two centuries is reviewed in Clifford J. Cunningham, *The Origin of the Asteroids: Olbers vs Regner* (Ft. Lauderdale, FL: Star Lab Press, 2012).

47 A. Crocker, *The universe; a philosophical poem* (Taunton: J. Poole, 1808).

28   Classical Deities in Astronomy

> Beyond the orb of Mars behold we find
> Four smaller bodies of the planet kind.-
> The first, though last reveal'd to human sight,
> Is Vesta call'd; of feeble, dusky light;
> Whose bulk and distance are to us unknown,
> Nor have her revolutions yet been shown.
>   Still farther off (with telescopic eye)
> The late discovered Ceres we descry;
> Of size minute, and various in her hue,
> Sometimes a red, at others, white or blue.
>   See Pallas, gliding on in annual round,
> The minimus of planet-stars, is found;
> Of size so small, as well as feeble light,
> No wonder she so long escap'd our sight.
>   In path elliptic, Juno wings her way,
> And feebly sheds on us her silver ray;-
> Her length of days (as yet to us unknown)
> By future observations, will be shown.
> Whate'er her bulk, her days how short or long,
> Creative judgment has not made them wrong.
> In ev'ry world, in ev'ry part, we find
> Th' unerring wisdom of th' eternal mind.

The mention of various colours being attributed to Ceres is also unique to this poem. Different observers discerned Ceres in different ways. The general consensus of modern telescopic observers is that Ceres appears white or bluish-white. Whether Herschel's observation of Ceres as reddish 'was a purely subjective problem, a physiological one, or down to his speculum metal being a better reflector at the long-wavelength end of the spectrum, is still open to debate'.[48]

8. Legacy

As a plethora of discoveries of more asteroids followed from the mid–nineteenth century on, the impulse to commemorate astronomical discoveries in verse evaporated, although the asteroids and 'The Georgian Planet' continued to be featured in prose. Among these is a poem by the

---

48 R. Holmes, *The Age of Wonder* (New York: Pantheon Books, 2008), p. 87.

American Lydia Sigourney, an anonymous poem entitled 'Immortality' from an English writer in 1839, and this example by Thomas Edgar that weaves in the purely classical attributes of the deities: Ceres, goddess of Agriculture, Pallas, who bestowed arts and science on humanity, Juno, the jealous wife of Jupiter, and Vesta, the goddess of the hearth and home.[49]

> Next, four twin sisters, lately known,
> In noble splendor do roll on—
> Ceres, who agriculture taught-
> Pallas, who arts and science brought
> To ancient Greece, as poets tell,
> In which she did the world excell—
> Juno, the watchful, jealous wife,
> Vesta, who virgin was for life.

The rich and fascinating verse elicited by the amazing astronomical discoveries of the late eighteenth and early nineteenth centuries, collected here for the first time, offers a unique insight into the golden age of planetary science.

---

49 Lydia Howard Sigourney, *Poems* (Boston: S. G. Goodrich, 1827); Anonymous, *Immortality, a Poem in Six Books* (London: John Hearne, 1829); Thomas Edgar, *Poems on various subjects*. Verse quoted from the poem 'On a Beautiful Aurora Borealis, which appeared in Autumn 1819, With a cursory sketch of the Heavens at the time', (Dumfries: J. M'Diarmid, 1822).

A Reinvestigation Into Astronomical Motifs in Eddic Poetry, with Particular Reference to Óðinn's Encounters with Two Giantesses: Billings Mær and Gunnlöð

Dorian Knight

Abstract: Archaeological and literary scholarship on Old Norse/Icelandic astronomy has been very scarce. This is no doubt due to what has been perceived as a lack of transparent evidence. Within this article I will re-evaluate how astronomical knowledge can be uncovered in Old Norse/Icelandic literary sources by analyzing mythologically problematic stanzas within Eddic Poem *Hávamál*.

Stanzas 96–110 of *Hávamál* tell of a self-contained mythic narrative concerning the god Óðinn's attempted seduction of two giantesses, Billings Mær and Gunnlöð. Using the analytic tools of cultural astronomy and a wide range of philological, cultural and medieval and modern ethnographic evidence as well as insights into other areas of Old Icelandic poetry, I suggest the following: that the relevant stanzas of *Hávamál* contain two complementary love stories fundamentally connected to the movement of the waxing and the waning of the moon. We can therefore understand the kernel of this myth as a description of what could be seen in the night sky, although the story is couched in the terms of an aphoristic tale, at the centre of which men and women's deceitfulness towards each other is portrayed with a comic twist.

My conclusion is therefore that it may be possible to interpret certain myths in the *Poetic Edda* in the light of a technical language that encodes astronomical information through allegory. Therefore, although it may at first seem unusual to locate Old Norse gods in the real-world observations of the night sky, we can choose to read them as a way of describing what is going on in the sky using mythological language.

Introduction and Review of Past Scholarship
Unfortunately, literary scholarship on Old Norse celestial observations has for the most part been a black hole. This is no doubt due to what has been

perceived as a lack of transparent evidence. The textual record holds no obvious references to the stars as exemplified in Ancient Greek culture, such as Ptolemy's astrological treatise *Almagest*, or Cicero's sixth-century vision of the Milky Way as occurs in *The Dream of Scipio*.[1] Likewise, the archaeological record is sparse; the Old Norse world has preserved no permanent material astronomical constructions equivalent to the Egyptian pyramids, the English Stonehenge or similar edifices found throughout ancient Latin America.[2] The lack of clear evidence is not to say that such cultural artefacts never existed in ancient and medieval Scandinavia, but those which still do survive are 'to be found scattered over many different kinds of sources and are very fragmented'.[3] Within this paper I intend to re-evaluate the evidence for the existence of astronomical motifs in Old Icelandic Eddic poetry.[4] Seeing how the lack of obvious references has not stopped scholars from attempting to find those that do exist, there is a small, yet growing body of academic work dealing with the issue, which I shall provide an overview of in the following paragraph. I shall then go on to an analysis of how astronomical knowledge can be uncovered in Old Norse literary sources and help illuminate mythologically problematic stanzas of the Eddic poem *Hávamál*.

---

1 Ptolemy, *Almagest*, G. J. Toomer, ed., (Princeton, NJ: Princeton University Press, 1998); Cicero, *The Dream of Scipio*, Sally Davis, ed., (London: Longman, 1988).

2 For discussions on the astronomical background to these geographically diverse sites see: Miroslav Verner, *Pyramids: The Mystery, Culture and Science of Egypt's Great Monuments* (New York: Grove Press, 2001); Anthony Johnson, *Solving Stonehenge: The Key to an Ancient Enigma* (London: Thames and Hudson, 2008); Anthony Aveni, 'Archaeoastronomy in the Ancient Americas', *Journal of Archaeological Research*, Vol. 11, (2003): pp. 149–91.

3 Christian Etheridge, *A Systematic Reevaluation of the Sources of Old Norse Astronomy* (in press).

4 The theoretical discussions on astronomy exceed the concerns for this essay. As the term is problematic from a theoretical perspective I shall use a simple definition suggested by Mark Williams, *Fiery Shapes: Celestial Portents and Astrology in Ireland and Wales 700–1700* (Oxford: Oxford University Press, 2010), p. xix: 'astronomy pertains to the rotation of the heavens, the rising, setting and motion of the stars, or else to the reason the stars got their names'.

Scholarship on Old Norse astronomy crystallised in the nineteenth century. The most noticeable authors included Jacob Grimm and Richard Hinckley Allen.[5] Grimm was a great compiler of Germanic mythological material and Hinckley Allen adapted many of Grimm's findings regarding Scandinavia in his own work, which dealt with various world myths and folklore connected to stars. The early twentieth century saw a blossoming in the number of writers willing to engage with the topic.[6] However, the research in the later part of the century proved more fertile and two important publications saw the light of day. The first was de Santillana's and von Dechen's profound and impressive attempt to crack the code of archaic cosmological symbolism, which included a brief analysis of the Scandinavian myths primarily through an inquiry of the twelve houses of the gods in *Grímnismál* in light of the houses of the zodiac.[7] In 1994 the influential *Star Myths of the Vikings* was published, the first book-length investigation into the subject that still to the present day remains the only substantial study of its kind. In more recent years the subject of astronomy in Old Norse myths has attracted writers on the fringe of academia, viewing the welkin through runic symbolism as well as cosmological and sacred numerological symbolism etched into medieval Icelandic literature and the landscape.[8] Although many writers may have skimmed the topic, the more recent sober and pioneering works of notable scholars such as Etheridge, Sigurðsson, Ogier, Persson and Kuperjanov all underline the

---

5 Jacob Grimm, *Deutsche Mythologie* (Göttingen: Dietrich, 1835); Richard Hinckley Allen, *Star Names: Their Lore and Meaning* (New York: Dover Publications, 1899).

6 Nils Beckman and Kristian Kålund, *Alfræði íslenzk II: Rímtöl* (Copenhagen: Samfund til udgivelse af gammel nordisk Litteratur, 1908); Nils Beckman and Martin Nilsson, *Tidsregning* (Stockholm: A. Bonnier, 1934); Jan de Vries, *Altgermanische Religionsgeschichte* (Berlin-Leipzig: W. De Gruyter, 1935).

7 Giorgio de Santillana and Hertha von Dechen, *Hamlet's Mill: An Essay Investigating the Origins of Human Knowledge and its Transmission through Myth* (New Hampshire: Springer-Verlag, 1992).

8 Einar Pálsson, *Hvolfþak himins* (Reykjavík: Mímir, 1985); Einar Gunnar Birgisson, *Egyptian Influence and Sacred Geometry in Ancient and Medieval Scandinavia* (Reykjavík: Self Published, 2004); Peter Halldórsson, *The Measure of the Cosmos* (Reykjavík: Salka, 2007): N. Pennick, *Runic Astrology* (London: Holmes Publishing Group, 1995).

fact that there is still a great deal of work yet to be done regarding astronomy in Old Norse literature. It is their work that provides a departure point for this paper.[9]

In the following pages I shall attempt to uncover how a methodological model to finding astronomical knowledge within the field of Old Norse myth can be found. I shall limit my study to the anonymous *Poetic Edda* contained in the *Codex Regius*, as it remains the most relevant Old Norse source for my essay as will be explained below. However, as Ogier points out, 'cultural astronomy as a field is painfully aware of the conjectural nature of its results'.[10] Furthermore, the inconsistent nature of the cultural record results in the evidence being based on scattered clues. Great methodological caution is therefore required and conclusions remain conjectural for the time being.

Historical Background to Astronomy in Old Norse Poetry
The study of the sky is a common activity across the globe. People from different cultures and times have always had to make sense of what was seen in the heavens above them and have used stargazing for a wide variety of purposes. Throughout world cultures the movements of celestial objects have helped determine timekeeping, explain terrestrial events and guide in matters of 'hunting, navigating and planting and to determine principles of leadership and community'.[11] It has also influenced more mundane matters, such as the most auspicious time to have one's hair cut. Therefore, due to the profound effect the planets had on diverse world cultures, it is inevitable that scholars would attempt to find traces of

---

9 Etheridge, *Old Norse Astronomy*, (in press); Gísli Sigurðsson, 'Goðsögur Snorra-Eddu: Lýsing á raunheimi með aðferðum sjónhverfingarinnar', *Rannsóknir í félagsvísindum X* (2009): pp. 851–61; James Ogier, 'Islands and Skylands: An Eddic Geography', in *Islands and Cities in Medieval Myth, Literature and History*, Andrea Grafetstätter et al., eds., (Frankfurt am Main: Peter Lang, 2009) pp. 9–12; Jonas Persson, '*Norse Constellations*', (2010). [Accessed 13 May 2012]. Available at: http://www.digitaliseducation.com/resources-norse.html; Andres Kuperjanov, '*Pseudomythological Constellation Maps*' (2006) [Accessed 20 August 2012]. Available at: http://www.folklore.ee/folklore/vol32/cps.pdf

10 Ogier, *Eddic Constellations*.

11 E. C. Krupp, 'Myths and their Interpretation', *Journal for the History of Astronomy*, Vol. 28, (1997): pp. 353–54.

astronomical interest in the Old Norse cultural record. Contemporary to the time when the Eddic poems were set down on vellum, astronomy was important across Europe and underlined a great deal to the medieval mind, from prediction of the future to the assessment of character. Likewise, the zodiac was carved in cathedrals, written in manuscripts and painted on walls all over Europe. This classical system inspired by Greek and Roman culture may have been brought to Scandinavia through the works of the Danish astronomer Petrus de Dacia, who bridging the gap between Denmark and France and through his collection of encyclopaedias entitled *Rimtöl* created 'a hitherto unknown link between Scandinavian astronomy and European science'.[12] These books describe the zodiac, planetary names and constellations, heavily influenced by Greek and Arabic sources from scholars such as Bede and Macrobius. Likewise, classical learning brought to Iceland through France and England would have contributed to knowledge of this subject in medieval Iceland. However, as Ogier has pointed out:[13]

> astronomical lore also derives from an older, inherited culture. The Scandinavians inherited from their Germanic, and, indeed, Indo-European stock the astronomical lore the descendents of which appear in, e.g., Vedi and classical Mediterranean mythology... One may assume a transmission, from generation to generation, of Indo-European mythologems through the agency of a metaphor who used the skies as his text.

It therefore seems possible to assume that myths relating to the sky must have entered or already been present in the Old Norse mythological system prior to Christianization due to the basic human necessity to perceive order in celestial patterns. This provides the fundamentals for cultural phenomena such as myths. It is the supposition that this inherited Indo-European knowledge survived and shines through in certain Old Norse myths, which will be examined in the next section.

---

12 Etheridge, *Old Norse Astronomy*, (in press).

13 Ogier, 'Islands and Skylands', p. 11.

The Sources: Eddic Poetry, a Definition of Myths and Ways of Reading Them in Light of Astronomy

The Old Norse myths come down to the modern reader (as previously mentioned) in Eddic poetry. Eddic poetry refers primarily to the anonymous poetic contents of the *Codex Regius* manuscript, namely 'anonymously transmitted poems that deal with the myths or heroic world of the Nordic countries'.[14] This poetic material is of varying age and origin. Although written down in the medieval period, the poems are essentially 'folk material, drawn from early Scandinavian oral tradition which at some stage seems to have adapted the poetic form'.[15] They 'bear the hallmarks of oral-traditional verse: it is alliterative, as are older analogous forms of Germanic oral poetry...and it is formulaic...which is characteristic of poetry originally composed and disseminated in a pre-literate culture'.[16] As such, Eddic poetry can be viewed as a 'fossil of a once living vital tradition'.[17] Seeing how the poems had roots in a pagan world of ideas (despite the fact they extended into the Christian era), they are likely to contain certain elements of pre-Christian astronomical traditions as mentioned above, far older than the manuscript in which they are actually preserved in. Using this assumption I shall now examine how astronomical knowledge may be encoded in Eddic poems, specifically in *Hávamál*.

On first glance, many poems in the *Poetic Edda* (including the relevant stanzas of *Hávamál*) seem to have little to do with astronomy. On this point, Hallberg claims that 'the demarcation of the imagery of the *Poetic Edda* is complicated for various reasons. Thus the very character of mythological poetry may make the identification of its metaphorical

---

14 Terry Gunnell, 'Eddic Poetry', in Rory McTurk, ed., *A Companion to Old Norse Icelandic Literature and Culture* (London: Blackwell Publishing Ltd, 2005), pp. 82–101.

15 Ibid., p.87.

16 Chris Abram, *Myths of the Pagan North—The Gods of the Northmen* (London: Continuum, 2011).

17 Ibid., p. 19.

elements problematic'.[18] The celestial elements in the Eddic poem *Völuspá* are a good example of this problem; Stanza 5 in English translation reads:

> From the South, Sun, companion of the moon,
> threw her right hand around the edge of the heaven;
> Sun did not know where her hall might be,
> The moon did not know what power he had.[19]

Although the exact meaning of the passage is unclear, the sun is being personified as a living being. As Hallberg points out regarding this passage, authorial intent is very hard to gauge for the modern reader; 'did he [the author] regard his work as representing authentic myth through and through, or was much of it, even to him, more or less metaphorical?'[20] As the myths are refracted through literature, which has likely removed them from any possible religious contexts they may once have had, it is 'only through the revealing of and definition of mythological models in the sources, that one may have a hope of coming close to genuinely pagan beliefs'.[21] I suggest that it is necessary to interpret these myths by looking at the 'underlying structure in the myth behind the details…to read the mythology much like one would read a parable. In a parable, it is the

---

18 Peter Hallberg, 'Elements of Imagery in the Poetic Edda', in Robert James Glendinnning & Haraldur Bessason, eds., *The Edda: A Collection of Essays* (Winnipeg: University of Manitoba Press, 1983), pp. 47–86.

19 'Sól varp sunnan,
   sinni Mána,
   hendi inni hægri
   um himinjódýr.
   Sól það né vissi,
   hvar hún sali átti,
   stjörnur það né vissu
   hvar þær staði áttu
   Máni það né vissi,
   Hvað hann megins átti'.

Gísli Sigurðsson, ed., *Eddukvæði* (Reykjavík: Mál og Menning, 1998), p. 6.

20 Hallberg, 'Elements of Imagery', p. 49.

21 Maria Kvilhaug, *The Maiden with the Mead: A Goddess of Initiation Rituals in Old Norse Mythology* (VDM Verlag, 2007), p. 19.

structure that carries the meaning, not what fills out the structure'.[22] My own perspective is similar to this in that in order to uncover the astronomical tradition in Old Norse literature, it is the structure of the myth that must be examined (although the details contained within the myth may of course also relate to the structure). I shall use this methodological model to examine a myth in the Eddic poem *Hávamál*. I have picked relevant stanzas of this poem, as I believe that they reflect knowledge of stargazing, specifically regarding the movement of the moon through the lunar cycle. This approach is explained more fully below.

An Astronomical Motif in *Hávamál*

Hávamál is the longest extant Eddic poem at 1087 lines, and arguably the most disparate. According to many literary-minded scholars, it is most likely a composite of (some) pre-existing oral poetic material and Latinate writing, and probably a conflation of pre-existing poems. There is however a great deal of uncertainty regarding the exact relationship of these pre-existing poems to *Hávamál* as it appears in the *Codex Regius*. According to Abram, this problem is due to the fact that

> we do not know whether their amalgamation into a single text occurred at a relatively early stage of transmission; or if the current form of the poem owes itself to the decision of an 'editor' who noticed the correspondences between the component poems and wove them together as part of the process by which they were written down.[23]

In 1891 the German scholar Karl Müllenhoff suggested that *Hávamál* is a sequence of six separate poems with the unifying theme of Óðinn as the narrator.[24] Most scholars have maintained this belief since Müllenhoff's time. In tandem with this methodology, a separate narrative considered as constituting one of these six separate poems is often entitled 'the Poem of

---

22 Ibid., p. 20.

23 Abram, *Pagan Myths*, p. 223.

24 Karl Müllenhof, *Deutsche Altertumskunde* (Berlin, 1883).

Sexual Intrigue' (this is explored in greater detail below).²⁵ I shall maintain this methodology throughout this essay, despite that the division of the poem in this manner remains theoretical.²⁶

As previously stated, the motif in question that I am dealing with concerns Óðinn's attempted seductions of the giantesses Billings Mær and Gunnlöð.²⁷ The Old Norse prose in English translation reads:

> 96 - That I found when I sat among the reeds
> and waited for my desire;
> body and soul the wise girl was to me,
> nevertheless I didn't win her.
>
> 97 - Billing's girl I found her on the bed,
> sleeping, sun radiant;
> the pleasures of a noble were nothing to me,
> except to live with that body.
>
> 98 - 'At evening, Odin, you should come again,
> if you want to woo yourself a girl;
> all is lost if anyone knows
> of such shame together.'
>
> 99 - Back I turned, and thought I was going to love,
> back from my certain pleasure;
> that I thought that I would have,
> all her heart and her love—play.

---

25 Although the two narratives (Óðinn and Gunnlöð and Óðinn and Billings Mær) are traditionally assumed to be connected to each other, it is a matter of debate as to where the two motifs both begin and end. To my mind the core of the two myths seem to be contained in stanzas 96–102 (Óðinn and Gunnlöð) and 104–110 (Óðinn and Billings Mær) as this is where the narrative shifts into the first person. However, other stanzas may refer to the myth, for example stanza 84.

26 It should be pointed out that there are several reasons for dividing the poem in this way. This includes that the two stories neatly pivot around the single theme of sexual treachery. Furthermore, the two narratives contain specialized vocabulary that exists nowhere else in *Hávamál*.

27 Normalised spellings of Old Norse names will be maintained throughout this essay.

100 - So I came afterwards, but standing ready
were all the warriors, awake,
with burning torches and carrying brands:
thus the path of desire was determined for me.

101 - And near morning when I came again,
then the hall company were asleep;
a bitch I found then tied on the bed
of that good woman.

102 - Many a good girl when you know her better
is fickle of heart towards men;
I found that out when I tried to seduce
That sagacious woman into shame;
Every sort of humiliation the clever woman devised for me,
And I didn't even possess the woman.

104 - I visited the old giant, now I've come back,
didn't get much there from being silent;
with many words I spoke to my advantage
in Suttung's hall.

105 - Gunnlöð gave me from her golden throne
a drink of the precious mead;
a poor reward I let her have in return,
for her open-heartedness,
for her heavy spirit.

106 - With the mouth of the auger I made space for myself
and gnawed through the stone;
over me and under me went the path of the giants,
thus I risked my head.

107 - The cheaply bought beauty I made good use of,
the wise lack for little;
for Óðrerir has now come up
to the rim of the sanctuaries of men.

108 - I am in doubt as to whether I would have come
back from the court's of giants,

If I had not made use of Gunnlöð, that good woman,
and put my arms about her.

109 - The next day the frost-giant went
To ask of the High One's advice, in the High One's hall;
They asked about Bolverk: whether he was amongst the gods,
Or whether Suttung had slaughtered him.

110 - I thought Odin had sworn a sacred ring-oath,
how can his word be trusted!
He left Suttung betrayed at the feast
and made Gunnlöð weep'. (Larrington, *Poetic Edda*, p.26-29) [28]

---

28 '96 - þat ec þá reynda,
er ec í reyri sat,
oc vœttak míns munar;
hold oc hiarta
var mér in horsca mœr;
þeygi ec hana at heldr hefic.

97 - Billings mey
ec fann beðiom á,
sólhvíta, sofa;
iarls ynði
þótti mér ecci vera
nema við þat líc at lifa

98 - Auc nær apni
scaltu, Óðinn, koma,
ef þú vilt þér mæla man;
alt ero ósköp,
nema einir viti
slícan löst saman.

99 - Aptr ec hvarf
oc unna þóttumz
vísom vilja frá;
hitt ec hugða,
at ec hafa mynda
geð hennar alt oc gaman.

100 - Svá kom ec nœst, at in nýta var
vígdrót öll um vakin;

með brennandom liósom oc bornom viði,
svá var mér vilstígr of vitaðr.
101 - Oc nœr morni, er ec var enn um kominn,
þá var saldrót um sofin;
grey eitt ec þá fann innar góðo knon
bundit beðiom á.

102 - Mörg er góð mœr, ef gorva kannar,
hugbrigð við hali;
þá ec þat reynda, er iþ ráðspaca
teygða ec á flœrðir flióð;
háðungar hverrar leitaði mér it horsca man,
oc hafða ec þess vœtki vífs.

104 - Inn aldna iötun ec sótta,
nú em ec aptr kominn,
fát gat ec þegiandi þar;
mörgom orðom
mælta ec í minn frama
í Suttungrs sölom.

105 - Gunnlöð mér um gaf
gullnum stóli á
drycc ins dýra miaðar;
ill iðgiöld
lét ec hana eptir hafa
síns ins heila hugar,
síns ins svára sefa.

106 - Rata munn
létomc rúms um fá
oc um griót gnaga;
yfir ok undir
stóðomc iötna vegir,
svá hætta ec höfði til.

107 - Vel keyptz litar
hefi ec vel notið,
fás er fróðum vant;
því þat Óðrerir
er nú upp kominn
á elda vés iaðar

My contention in this essay is that these two separate, yet interrelated narratives represent a semantic whole in which we are able to detect a basic meaning of a myth regarding the movement of the moon.[29] However, this basic meaning seems to be hidden under the theme of sexual treachery. As a recent analyst points out; 'the tone is light and sometimes self-mocking, with an air of cynical balance which implies that both men and women behave either as exploiters or as fools and that they are as bad as each other'.[30] However, the deeper meaning of the myth remains highly problematical. As Svava Jakobsdóttir points out regarding the story of Óðinn and Gunnlöð, 'few recent scholars have dealt with the obscure and

---

108 - Ifi er mér á,
at ec væra enn kominn
iötna görðum ór,
ef ec Gunnlaðar né nytac,
innar góðo kono,
þeirar er lögðomc arm yfir

109 - Ins hindra dags
gengo Hrímþursar,
Háva ráðs at fregna,
Háva höllo í;
at Bölverki þeir spurðo,
ef hann væri með böndum kominn
eða hefði hánom Suttungr of sóit.

110 - Baugeið Óðinn,
hygg ec at unnit hafi;
hvat skal hans trygðom trúa?
Suttungr svikinn
hann lét sumbli frá
oc grætta Gunnlöðo.'

Gustav Nekel and Hans Kuhn, eds., *Edda: Die Lieder des Codex Regius nebst verwanten Denkmälern* (Heidelberg: Winter, 1962), pp. 33–34.

29 As I pointed out above, naked eye astronomy and a wider Indo-European background provide likelihood for the appearance of astronomical phenomena in Old Norse mythology, which would have naturally included the moon.

30 John McKinnell, 'Hávamál B: A Poem of Sexual Intrigue', *Saga Book of the Viking Society*, Vol. 20, (2005): pp. 83–114.

44 A Reinvestigation Into Astronomical Motifs in Eddic Poetry

difficult mythological elements'.[31] Neither has the narrative of Óðinn and Billings Mær received much more attention in secondary literature. However, most academics, such as Margret Clunies Ross have agreed that the story of Óðinn's failed seduction of Billings Mær is often contrasted with the successful seduction of Gunnlöð in the prior episode.[32] Noticeably, the majority of these commentators have had relatively little to say about the mythological implications of these stanzas. My suggestion is that the story can be split into two halves, the first half (Óðinn and Billings Mær) formulated around the moon in decline as the verses progress, the second half (Óðinn and Gunnlöð) associated with its increase, and ending with the fear that the moon may disappear again, thus coming full circle. The myth here can therefore be arguably read as a technical language that encodes astronomical information through allegory.

Óðinn and Billings Mær
I suggest that it is necessary to cast the net into the periphery of the myth by examining stanza 84 in *Hávamál*, as this may inform the rest of the narrative.[33] The verse in translation reads:[34]

> The words of a girl no one should trust,

---

31 Svava Jakobsdóttir, 'Gunnlöð and the Precious Mead', in Caroyne Larrington and Paul L. Acker, eds., *The Poetic Edda: Essays on Old Norse Mythology* (London: Routledge, 2007), pp. 27–58.

32 Margret Clunies Ross, *Prolonged Echoes: vol.1* (Odense: University Press of Southern Denmark, 1994).

33 McKinnell (Hávamál, pp. 83–114) suggests that stanza 84 acts as an introduction to the story of Óðinn and Billings Mær and is a warning of female unreliability towards men that is obvious in Billings Mær's deception of Óðinn. I shall maintain McKinnell's methodology that this stanza is of relevance to the myth of Óðinn and the daughter of Billingr. See footnote 10.

34 'Meyiar orðom,
  scyli mangi trúa,
  né því er qveðr kona;
  þvíat á hverfanda hvéli
  vóro þeim hiörto scöpuð,
  brigð í brióst um lagit' (Nekel and Kuhn, *Edda*, p. 30)

nor what a woman says; for on a whirling wheel their hearts were made
deceit lodged in their breasts.[35]

The phrase 'a whirling wheel' (*hverfandi hvél*) is mythologically problematic. In the form in which it appears in *Hávamál* it has invariably been translated as referring to a lathe;[36] as demonstrating medieval Christian influence from the Old Testament's apocryphal Ecclesiasticus;[37] or a potter's wheel.[38] The phrase appears elsewhere in the Old Norse literary corpus; for example in *Óláfs saga Tryggvasonar* (chapter 67) in *Flateyjarbók*, in which it is associated with fortune, possibly under the influence of Boethius's *Consolation of Philosophy*.[39] However, another interpretation is possible, one which may reflect stargazing. In the Eddic poem *Alvíssmál* (stanza 16), *hvél* is incorporated into the compound *fagrahvél*, referring to the sun. The word however may also refer to the moon. In *Alvíssmál* (stanza 14) the moon is referred to as *hverfanda hvél*, whilst in the anonymous *Líknarbraut* the moon is called *hvel mána*.[40] In relation to the phrase as it appears in *Hávamál* Kristján Albertsson has suggested that it may refer to a certain likening of women's moods and the menstrual cycle to the phases of the moon.[41] In his own words:

> In people's eyes the moon was always a magical, mysterious celestial body which could make the sea swell

---

35 Larrington, *Edda*, p. 25.

36 Carolyne Larrington, '*Hávamál* and Sources outside Scandinavia', *Saga Book of the Viking Society*, Vol. 23, (1991): p. 148.

37 Nore Hagmen, 'Kring några motiv i *Hávamál*', *Arkiv för filologi*, Vol. 72, (1957): p. 13.

38 David A. Evans, 'More Common Sense About *Hávamál*,' *Skandinavistik*, Vol. 19, (1989): p. 115–16.

39 See McKinnell, *Hávamál*, p. 96.

40 See Ernest Albin Kock, *Den norsk-isländska skaldediktningen* I–II (Lund: C. W. K. Gleerup, 1946–49.)

41 Kristján Albertsson, 'Hverfanda hvel', *Skírnir*, Vol. 151, (1977): pp. 57–58.

> or dwindle from the beaches, but also influence the mood and well-being of people, and not necessarily for the better...both the moon and the woman had an equally long month which leads to an assumption about an extraordinary connection, a sort of blood bond—and about the power the moon has over the moods of women...to him [the editor], their [women's] emotional life seemed to resemble nothing else but this celestial body.[42]

What this suggests is that the editor of these stanzas in *Hávamál* connected the imagery of the moon with women, most obviously in the movement of the moon across the night sky and its connection and possible responsibility for the frequently changing moods of women.[43] It seems likely that although the phrase may have had other meanings as well it was also used in this lunar context.

Although fleeting references to the moon in Old Norse exist, such as that analysed in the preceding paragraph, full-length narratives are rare. Indeed the only one is that found in *Snorra-Edda*.[44] However, lack of clear

---

42 'En í augum manna var tunglið frá aldaöðli töfrandi, dularfullt himinhvel, sem gat látið haf hœkka og hníga fyrir ströndum, en líka orkað á geð og líðan manna, og ekki œfinlega til góðs... Tungl og kona höfðu jafn-langan mánuð, en af því óx grunar um kynleg tengsl, einskonar blóðbönd—og um vald tungls á skaplyndi kvenna... Honum hefur þótt sem tilfinnigalíf þeirra minnti ekki á annað fremur en einmitt þetta himinhvel' (my translation).

43 In general, the moon was considered to be affecting human moods. For example, the English word 'lunatic', which is derived from the Latin *lunaticus* ('struck by the moon'), expresses the notion that periodic insanity was the result of the moon's periodic changes. Likewise, the Icelandic word *tunglsjúkur* (literally 'moon sick') can be translated as insane or epileptic. As Leonhard Franz (in 'Die Geschichten vom Monde in der Snorra-Edda', *Mitteilungen der Islandfreude X* (1922/23), pp. 45–49) pointed out, the notion that the connection between the moon and women is particularly strong lies close at hand. The fertility cycle of women corresponds to the lunar month (28 days), so that the Latin word *menstruus* (month), derivative of *mensis* was coined.

44 See Anne Holtsmark, 'Bil og Hjuke', *Maal og Minne* (1945), pp. 139–54, for an account of Máni and Sól in *Gylfaginning*. Many scholars turn to *Snorra-Edda* as a first attempt to explore Old Norse myths. Snorri was the first to synthesize the myths into a written narrative. He was also essentially a well-educated Christian, writing about and shaping the Old Norse myths in the light of Christian universal

evidence does not reflect the idea that there was only one tradition of belief about the moon. Indeed there were most likely more that have been lost to modern scholarship. An alternative way of defining and thinking about the moon is arguably indicated by a kenning preserved in *Hákonardrápa*, composed in the tenth century on behalf of King Hákon by Guðormr Sindri, which presents the moon as arguably being a person in possession of a woman.[45] Furthermore, in a more recent Icelandic fairytale the moon is euphemised as a king who has a beautiful daughter.[46] This indicates a possible other tradition of thinking about the moon that has for the most part been lost to us. This idea of the moon as being a man in some way possessing a woman must be born in mind for later.

The moon is actually referenced elsewhere in *Hávamál*, in stanza 137. Although this stanza is not connected to the myth of Óðinn and the attempted seduction of the two giantesses, it may be useful to understand how people thought of the moon. The stanza in translation reads:[47]

---

history and to suit the needs of a Christian elite literary audience. However, his accounts should not be viewed as having a monopoly on Old Norse mythic narratives, and there were likely aspects of Old Norse myths not mentioned in *Snorra-Edda*.

45 Composed in King Hákon's reign, as Abrams ('Pagan Myths', p. 101) points out, this poem was composed 'in an area where Norse mythology was still very much a living force' in the tenth century.

46 In this tale the King Máni has a daughter by the name of Mjaðveig, whose name means mead strength. This has interesting implications as the moon is connected to a mead container in the later stanzas of *Hávamál*. See: Adeline Rittershaus, *Die neuisländischen Volksmärchen* (Halle: Max Niemeyer, 1902).

47 'Ráðumc þér, Loddfáfnir,
    enn þú ráð nemir,
    nióta mundo, ef þú nemr,
    þér muno góð ef þú getr:
    hvars þú öl dreccir,
    kjós þú þér iarðar megin,
    þvíat iörð tecr við ölðri,
    enn eldr við sóttom,
    eic við abbindi,
    ax við fjólkyngi,
    höll við hýrógi
    heiptum skal mána qveðia

> I advise you, Loddfafnir, to take this advice,
> It will be useful if you learn it,
> Do you good if you have it:
> Where you drink ale, choose the power of earth!
> For earth is good against drunkenness, and fire against sickness,
> oak against constipation, an ear of corn against witchcraft,
> the hall against household strife, for hatred the moon
>   should be invoked
> earthworms for a bite or sting, and runes against evil;
> soil you should use against a flood.[48]

The second half of this stanza is evidently concerned with a list of medical remedies. As Evans has illustrated, some of this advice no doubt has a great deal of antiquity.[49] For example, regarding the phrase *jarðar megin*

> there may well be a specific connection with the so-called *terra sigillata,* cakes of earth rich in iron oxide, stamped with the image of Diana or Christ, exported from Lemnos and recommended (e.g. by Pliny and Galen) as a remedy against poison…there may also be a connection with the more general belief in the earth's holy and curative properties[50]

Regarding the phrase *heiptum skal mána qveðia*, this means that in order to get rid of hatreds (whatever they may be), one must call upon the moon. Therefore the moon can be seemingly viewed as having a positive effect in some sense in this line from *Hávamál*. It may be the case as suggested by Evans that the 'hatreds' referred to are 'the workings of the evil eye,

---

  beiti við bitsóttum,
  enn við bölvi rúnar;
  fold skal við flóði taka' (Nekel and Kuhn, *Edda*, pp. 39–40).

48 Translation by Larrington, *Poetic Edda*, p. 34.

49 Evans, *Hávamál*, pp. 130–34.

50 Ibid., p. 131.

against which moon-shaped amulets were employed in classical antiquity'.[51]

Bearing in mind that fragmentary evidence suggests that the moon had a positive effect it is worthwhile examining this in greater detail. As noted by Klingenberg, 'according to the widespread sympathetic belief, the earthly existence and human activity are bound to the visible increasing and decreasing of the moon, from Pliny the Elder to Paracelsus and homeopathic cures'.[52] He highlights one (of countless) folk beliefs that conforms to this idea:[53]

> Everything that is supposed to grow, to become good and big, everything that is supposed to develop into fertility, has to be begun and done as long as the moon is growing; because everything done along the with the growth of such a big celestial light will proceed well; but everything that is supposed to vanish, wither, die and be exterminated, that is the annihilation of all the disgusting and evil, has to happen when the light is decreasing to vanish completely.[54]

---

51 Ibid.

52 'Nach weit verbreitetem Sympathieglauben sind irdisches Sein und menschliches Tun an das sichtbare Werden und Schwinden des Mondes gebunden, von Plinius dem Alten bis Paracelsus und darüber hinaus Sympathiekuren bekannt' (my translation).

53 Heinz Klingenberg, 'Hávamál. Bedeutungs und Gesaltwechsel eines Motivs', in Oskar Bandle et al., eds., *Festschrift für Siegfried Gutenbrunner zum 65. Geburtstag um 26. Mai 1971. Überreicht von seinem Freunden und Kollegen* (Heidelberg: Winter, 1972), pp. 117–44. This is an extremely wide ranging phenomenon, and it would be surprising and unusual if such a belief did not also occur in the Old Norse world. As Robert Means Lawrence (in *The Magic of the Horseshoe: With Other Folklore Notes*, 2nd ed., [Forgotten Books, 2008], pp.15–16) points out 'the moon has always been considered the most influential of the heavenly bodies…the alleged prominent influence of the moon's wax and wane over the growth and welfare of vegetation was formally generally recognized'. He cites a wide variety of folklore, from medieval England to modern America.

54 'Was wachsen, gut und groß werden, was irgendwie eine fruchtbare Weiterentwicklung gewinnen soll, muß begonnen und getan werden, solange der Mond von der Sichel bis zur Vollendung seiner Füller wächst; denn was

I suggest that in light of this folk belief it may be the case that Óðinn's unsuccessful courting of Billings Mær in the first episode was due to it being set in the time of a waning moon; whilst Óðinn's successful courting of Gunnlöð was due to its setting in the favourable time of a waxing moon.

An analysis of the personal names and descriptions as found in the Billings Mær episode may also inform the theory that the episode is concerned with the waning of the moon. Klingenberg points out that 'Billingr derives from an iterative form of the Old Norse *bila* (to give in, to become slack and fail) and *bil* 'stay, time, movement, weak point', which is connected to the name of the goddess *Bil*'.[55]

A final point may tie the story to a mythical allegory about the moon. The penultimate line of the final stanza points to Óðinn finding a female dog substituting for Billings Mær, much to his disappointment. Dogs were of great importance in the Old Norse world. As indicated by Gräslund across Old Iranian and Celtic cultures the dog has a high status, both mythically and in reality.[56] In Scandinavian sources as diverse as Adam of Bremen, Thiermar of Merseburg and Ibn Fadlan, there is a demonstrable connection between dogs and death. In her recent article, Gräslund has

---

gleichzeitig mit dem Wachstum eines so großen Himmelslichtes geschiet, geht leicht vonstatten; was aber vergen, verdorren, austerben und ausgerottet werden sol, also die Vernichtung alles Widrigen, Bösen, hat während der Abnahme des Lichtes bis zu seinem völligen Verschwinden zu geschehen' (my translation).

55 Klingenberg, *Hávamál*, p. 121. 'Abgeleitet von einer Iterativ-bildung zu altnord. *bila* "nachgeben, sclaff werden; fehlschlagen", *bil* "Aufenthalt; Zeit, Augenblick, schwade Stelle", wozu auch der Name einer Göttin *Bil* gehört' (my translation). Further implications that this mythic motif may have its roots in astronomical observation lie in the fact that the goddess Bil appears in *Gylfaginning* in *Snorra-Edda* as a character who is visible in the moon along with her brother Hjúki. Holtsmark (*Bil og Hjuke*) has suggested that the myth attached to this explains the moon's phases. Klingenberg (*Hávamál*, p. 119) goes on to suggests that Billingr may also be related to an older myth about the moon. Although the connection between the story of Bil and the narrative of Óðinn and the two giantesses is very unclear, there may be a connection as the personal names suggest.

56 Anne-Sofie Gräsland, 'Wolves, Serpents and Birds. Their symbolic meaning in Old Norse Belief', in Anders Andren, Kristina Jennbert, and Catharina Raudvere, eds., *Old Norse Religion in Long Term Perspectives: Origins, Changes and Interactions, an International Conference in Lund, Sweden, June $3^{rd}-7^{th}$, 2004* (Lund: Nordic University Press, 2006), pp. 124–29.

posited that 'the mythical dog seems to be a medium on the border between the living and the dead, and in all likelihood the archaeological material reflects this important symbolic mythological meaning in the transformation from life to death'.[57] Indeed the historian of religion Bruce Lincoln has seen connections between the dogs in *Grímnismál* and *Völuspá* and the Indo-European dog of the realm of the dead.[58] Similarly, the Irish scholar Kim McCone has pointed to the close connection between dogs/wolves and Indo-European warrior ideology.[59] In his opinion, dogs as sacrificial animals could be seen as substitutes for wolves. In regards to the moon, we know from certain Eddic poems including *Völuspá* and Snorri's *Gylfaginning* that the wolf Fenrir's offspring will catch up with and eat the moon. There was therefore a belief in the Old Norse world about a canine eating the moon, which sheds an interesting light on the appearance of a female dog in place of Billingr's daughter, which now seems to have an appropriate symbolic meaning in accordance with the above-mentioned motif.[60]

The previous fragmentary evidence leads to the following conclusion regarding the Billings Mær episode: in the Old Norse world there was a narrative that characterized the moon related to the concept of a man being in possession of a woman. Unfortunately, this story has mostly been lost to us in the surviving Old Norse corpus. However, in Óðinn's encounter with the two giantesses this idea may still shine through. Stanza 84, as a precursor to the story, illustrates the close relationship between women and the lunar cycle. In stanza 97 the moon is full and Óðinn watches the

---

57 Gräsland, 'Wolves', p. 124.

58 Bruce Lincoln, *Death, War and Sacrifice: Studies in Ideology and Practise* (Chicago: University of Chicago Press, 1991). p. 96.

59 Kim R. McCone, 'Hund, Wolf und Krieger bei den Indo-germanen', in Wolfgang Meid, ed., *Studien zum Indogerman-ischen Wortschatz* (Innsbruck: Institut für Sprachwissenschaft, 1987), p. 102.

60 Interestingly the word used to describe the dog in *Hávamál*, *grey*, is very uncommon in the Old Norse literary corpus. It primarily occurs connected to mythological wolves in Eddic poetry (*Helgakviða Hundingsbana I*, *Hamðismál*, *Skírnismál* and *Þrymskviða*. The only other association is in *Atlakviða*, where the word relates to female sexuality. In light of the lunar connections outlined in this essay the correlation to a mythological wolf who eats the moon is most likely.

shining (*sólhvíta*[61]) Billings Mær in the hope of soon possessing her. Unfortunately, he is prevented from doing so by stanza 100. By stanza 101 it is clear to the reader that Óðinn's seduction has failed and he is left unfulfilled. If his sexual possession of the giantess is connected to the lunar cycle, his failure to seduce may serve as an indication as to why the moon wanes and (seemingly) disappears prior to the new moon. The appearance of a dog tied to the bed in place of Billings Mær at the very end of the episode may be a play on the symbolic meaning that dogs had in the Old Norse world in relation to beliefs about monstrous wolves/dogs threatening the moon with eclipse.

As ethnographic evidence suggested above, the growth of the moon is reflected in the prosperity of vegetation and growth, whilst the inverse is also true which is shown in these stanzas with Billings Mær. I shall now look to the episode with Gunnlöð to see how this idea is further supported.

Óðinn and Gunnlöð

Positive connection between marriage and the moon occurs throughout world cultures. Klingenberg highlights examples of such folk beliefs, that institutionalized weddings should take place at specific times of the month in order to maximize happiness, namely weddings should be celebrated and marriage contracts made during the waxing moon.[62] In the episode with Gunnlöð it may be the case that a wedding takes place. This is suggested in the line *Ins hindra dags* from stanza 109, as the phrase is also found in medieval Norwegian and Swedish laws to designate the day after a wedding has taken place.[63] Unfortunately, our knowledge of what was involved in a pagan wedding in the Old Norse world is extremely limited.[64]

---

61 It has been suggested to me through personal communication with Gísli Sigurðsson that the Norse would have been familiar with the idea that the moon was a white colour due to the reflections of the sun, hence why Billings Mær may indeed be described as 'sun white' and yet still be associated with the moon.

62 Klingenberg, *Hávamál*, p. 20.

63 See Evans, *Hávamál*, p. 123.

64 See Jens Peter Schjødt, *Initiation Between Two Worlds: Structure and Symbolism in Pre-Christian Scandinavian Religion* (Aarhus: University Press of Southern Denmark, 2008) for an examination of the few myths and possible rituals

However, it seems possible to mention that the fertility aspect (namely the ability to have children and to continue the family) was likely very important.[65] Likewise a great range of ethnographic evidence (both medieval and modern) points to the moon as being of key importance in phases of agriculture, particularly regarding the waxing moon creating favourable conditions for the growth of vegetation. If we can assume that a connection between weddings and fertility occurred, and the waxing moon was very influential in both areas, this would strengthen the lunar connection in the relevant stanzas of *Hávamál* between Óðinn and Gunnlöð.

Personal names may also give a clue as to the astronomical kernel that lies behind the narrative. In *Snorra-Edda*, where the story of Óðinn and the two giantesses is re-told, the giant Gillingr appears instead of Suttungr (he does not appear in the *Hávamál* version). Arguments have however been put forward that the *Hávamál* myth is simply referential to the *Snorra-Edda* myth and therefore merely alludes to the details that are present in Snorri's account. Meulengracht Sørensen sums this up:[66]

> The myth cannot be understood on the basis of these stanzas alone. They presuppose that it [the myth] is known beforehand. Nor is it the main purpose of these passages in *Hávamál* to tell the myth....This kind of relationship between an extant poem about gods and the underlying myths is not uncommon. It can rather be said to be the rule. The Edda poem presupposes that the myth is known,

---

that have survived across the Old Norse mythological corpus connected to weddings.

65 As Adam of Bremen (in his 'Gesta Hammaburgensis Ecclesiae Pontificum', in Werner Trillmich, ed., *Quellen des 9. und 11. Jahrhunderts zur Geschichte der hamburgischen Kirche und des Reiches* (Darmstadt: Wissenschaftliche Buchgesellschaft, 1961, p. 27) points out, at a wedding people turn to Freyr with sacrifices. Before that we learn that Freyr has a phallus. Similarly in the Eddic poem Þrymskviða, Þórr's hammer may have taken on the role of a fertility symbol, and thus that of the giver of fertility. Both these examples lead Schjødt, *Initiation*, p. 334 to claim that the 'connection between marriage and fertility is clearly marked'.

66 Preben Meulengracht Sørensen, 'Om Eddadigtenes Alder', in Gro Steinsland, et al., eds., *Nordisk Hedendom: Et Symposium* (Odense, 1991), p. 223.

and it often only reproduces part of it and often in a specific connection.⁶⁷

Meulengracht Sørensen's comment is correct in that most of the mythological Eddic poems (including *Hávamál*) are 'enumerative' and essentially composed of myth catalogues as identified by Klingenberg:⁶⁸

> Myth catalogues assume an audience initiated into mythological lore. The deeds of the gods and other individual myths are compressed into highly terse utterances, abbreviations which have a referential function as they assume more extensive narration elsewhere of the myth to which they allude.

So the details of *Hávamál* were likely known in greater depth by the intended audience than the poem itself assumes. Therefore it is plausible that Gillingr occurred at some stage of this myth, although he is not mentioned explicitly in *Hávamál*.⁶⁹ The name Gillingr may be related to the Gothic word *gil*, meaning a sickle, thus potentially providing a connection to the shape of the moon. Likewise the name Suttungr (son of Gillingr) may reveal something very similar; Klingenberg has suggested that the name Suttungr has its origins in the Norwegian verb *sutta* (to move very fast and quickly) and northern Swedish *sutta* (to jump).⁷⁰ Interestingly, in the Eddic poem *Alvíssmál* (stanza 14) the moon is referred to as *skyndir* (the hurrying one), derivative from Old Norse *skynda* (to move forward very fast), *skunda sér* (to accelerate), with cognates in

---

67 'Myten kan ikke forstås alene på grundlag af disse strofer. De forudsætter altså, at den på forhånd er bekendt. Det synes heller ikke at være Hávamál-passagens vigtigste formål at fortælle myten... Denne slags relation mellem et bevaret gudedigt og bagvedliggende myte er ikke sjælden. Den må snarest siges at være reglen. Eddadigtet forudsætter myten bekendt, og gengiver ofte kun en del af den og tit I specielle sammenhænge' (my translation).

68 Klingenberg, *Hávamál*, p.135.

69 As Schjødt *Initiation*, p.160 points out; 'is it difficult to see how important the episode with Gillingr is. His presence in Snorri's narrative may have had a mythical significance that is hidden from us today'.

70 Klingenberg, *Hávamál*, p.135.

Norwegian *skynde sig* (to hurry), Anglo-Saxon *skundian* (to drive, to move forward) and Old High German *scunten*.[71] The name Gunnlöð may also have similar connotations; it is very rare in the Old Norse literary corpus and is thought to mean 'War-invitation'. As a means of explanation, this name and therefore likely the character of Gunnlöð herself could have roots in a phenomena that was attributed to the Germanic tribes in Tactitus's *Germania*.[72] The prose in English translation reads:[73]

> About minor matters the chiefs deliberate, about the more important the whole tribe. Yet even when the final decision rests with the people, the affair is always thoroughly discussed by the chiefs. They assemble, except in the case of a sudden emergency, on certain fixed days, either at new or at full moon; for this they consider the most auspicious season for the transaction of business.[74]

Likewise in Caesar's *De Bello Gallico*, it is stated that the Germans when they start a fight, do so before a new sickle of the moon.[75] This

---

71 See Klingenberg, *Hávamál* for further details on these etymologies.

72 Although Tacitus is writing over 1000 years prior to when Eddic poetry was written down and many things may have changed a lot during that period, his works may at times be used for supporting evidence. However, it must be kept in mind that Tacitus's reasons for writing his works rarely coincide with our reasons for reading them.

73 Tacitus, *Germania*, at:
http://www.sacred-texts.com/cla/tac/g01010.htm [accessed 4 Sept. 2012].

74 De minoribus rebus principes consultant; de majoribus omnes: ita tamen, ut ea quoque, quorum penes plebem arbitrium est, apud principes pertractentur. Coeunt, nisi quid fortuitum et subitum inciderit, certis diebus, cum aut inchoatur luna aut impletur: nam agendis rebus hoc auspicatissimum initium credunt. (original and translation found at:
http://www.sacred-texts.com/cla/tac/g01010.htm, [accessed 4 September 2012].
See bibliography entry under 'Tacitus'.

75 See Klingenberg, *Hávamál*, p. 121 for further details. See Julius Caesar, *The Gallic War: Seven Commentaries on the Gallic War with Eighth Commentary by Aulus Hirtius*, trans. Carolyn Hammond, (Oxford: Oxford University Press, 2008) for translations of the original.

corresponds to the idea that things turn out prosperously during a waxing moon. Indeed, the philological evidence regarding Gillingr's, Suttungr's and Gunnlöð's names all seem to correspond to semantic categories of growth, moving forward, fullness and prosperity in battle, all connotations associated with an increasing moon.

Another point may suggest the waxing moon occurring in this narrative. In Stanza 107 it is written that 'for Óðrerir has now come up/to the rim of the sanctuaries of men'. This has proved to be mythologically problematic.[76] In *Snorra-Edda* the name refers to one of the three containers in which the mead is stored by the giant Suttungr. In stanza 140 of *Hávamál* this also seems to be the case. It has been put forward that in stanza 107 the name refers to the mead itself. However, I shall maintain the view that it refers to the container for the sake of simplicity. This container of the mead, Óðrerir, may actually pertain to the face of the moon. Hinckley Allen writes that

> in southern Sweden, a brewing-kettle is imagined on the moon's face; in northern Germany and Iceland, Hjuki and Bil with their mead burden, the originals of our Jack and Jil with their pail of water, the contents scattered or retained according to the lunar phases.[77]

The fullness of the moon's face is therefore dictated by the varying fullness of the contents of a pail of water. In Björn Jónsson's mapping of interstellar configurations in conjunction with the Old Norse myths, Bygir (the well) is equated to the moon halo, whilst Sægur the container or tub is matched to the moon face itself.[78] If this is correct it would illustrate that the Norse associated the face of the moon with a full vessel of liquid. Although there are inevitable source problems in how this idea as manifested in *Snorra-Edda*, connected to the myth contained within *Hávamál*, the lines 'for Óðrerir has now come up/to the rim of the sanctuaries of men' could imply that the container Óðrerir (symbolically

---

76 See Evans, *Hávamál*, p.12 for a historiographical analysis of the word Óðrerir.

77 Hinckley Allen, *Star Names*, p. 267.

78 Björn Jónsson, *Star Myths*, pp. 186–87.

the full moon's face) has now risen to the rim of the sanctuaries of men.[79] This seems appropriate as a description of a waning moon gaining ascension and shining over man's kingdom.

We know from stanza 104, that despite Óðinn's success with Gunnlöð, he eventually abandons her, weeping. It is mentioned in stanza 110 that he breaks a ring oath. It occurs across the Old Norse literary corpus that the swearing on rings was commonplace.[80] As Hayeur Smith points out 'rings in general are by their nature charged with spiritual value. The very shape being circular and closing upon itself symbolizes unity'.[81] In *Landnámabók* it is stated that a ring was to lie on the altar of every chief's temple, and is to be worn by the *goði* at assemblies.[82] Ring oaths also likely occurred within marital contexts, which is appropriate for the story of Óðinn and Gunnlöð. If the lunar cycle is considered as relating to a man who possesses a woman, the betrayal that Óðinn incurs in stanza 110 breaks the previous unity between him and the giantess, and sets the mythic narrative up for repetition; Óðinn is left single, not in possession of a woman, meaning the moon can begin its cycle of decline followed by regrowth again. It would seem likely that to the mind-set of the Norse people that there was a great deal of fear in the thought that the moon when waning

---

79 The final line has not ever been solved successfully. As the variety of opinions and interpretations suggests, the line is intractable. See Evans, *Hávamál*, p. 122.

80 See Francis P. Magoun, 'On the Old-Germanic Altar- or Oath-Ring (*Stallahringr*)', *Acta Philologica Scandinavica* XX (1949): pp. 277–93 for instances where ring oaths are involved from a historical perspective. The literary and mythological evidence also portrays strong ideas connected to rings, for example, Óðinn's possession of the ring *Draupnir*. Medieval Icelandic archaeological evidence suggests that something similar occurred in Iceland as well. As Hayeur Smith (in *Draupnir's Sweat*, p. 88) points out 'Icelandic graves offer three examples of possible amuletic rings stemming from burial contexts. Two of these are made of twisted copper alloy wire…the third from Álaugarey [*sic*] may have acquired its amuletic properties from the material it was made of'.

81 Michèle Hayeur Smith, *Draupnir's Sweat and Mardöll's Tears: An Archaeology of Jewellery, Gender and Identity in Viking Age Iceland* (Oxford: John and Erika Hedges Ltd, 2004), p. 88.

82 Jakob Benediktsson, ed., *Landnámabók* (Reykjavík: Hið Íslenska Fornritafélag, 1968), pp. 313–15.

would never return to its previous fully illuminated state.[83] As Franz points out, the idea

> that the shining lunar globe would be destroyed is widely documented in the myths of the Indo-Germanic peoples and even beyond. The dark side of the decreasing moon is continuously observed (hereby, the pictorial interpretation of the moon phases is always connected to the idea of a battle between a good (bright) and bad (dark) power.[84]

Within a specifically Old Norse context, we know from *Völuspá* stanza 40 that a portent of the end of the world is a giant wolf eating the moon.[85] This fear about the disappearance of the moon illuminates what Óðinn's betrayal of the giantess Gunnlöð may have signified: the belief that once the moon had reached its full brightness and started to wane, it may never return to its full luminescence again. The movement of the moon and its

---

83 A great deal of ethnographic evidence, both medieval and modern across western Europe supports this idea. A homily penned by the Frankish Benedictine monk Rabanus Maurus Magnentius (c. 780–856) claims that noise created during lunar eclipses was designed to scare away monsters threatening the moon. Likewise the Frankish king Charlemagne declared that during the moon's eclipse, a moon worm intends to eat the moon and must be persuaded not to, by men's screams. As late as the sixteenth century the German satirist and publicist Johann Fischart mocked this superstition.

84 Leonhard Franz, 'Die Geschichten', p. 49: 'im Mythos der indogermanische Völker und darüber hinaus oft belegte Vorstellung, daß die leuchtende Mondscheibe von einem dunklen Dämon vernichtet wird. Als solcher wird dann regelmäßig der dunkle Teil der abnehmenden Scheibe betrachtet (wobei mit dieser bildhaften Ausdeutung der Mondphasen stets auch die die Vorstellung vom Kampfe zwischen einer guten (hellen) und bösen (dunklen) Macht verbuden ist' (my translation).

85 In the pertinent stanzas of *Völuspá*, the wolf Mánagarmr is referred to as the 'robber of the *tungl*'. The word *tungl* has been interpreted by some as the sun. However, the word, also having the general meaning of 'heavenly body', refers to the moon. In Gylfaginning the *sól* and *tungl* (sun and moon) are differentiated. Likewise in Skáldskaparmál there is a contrast between *sól*, *tungl* and *himintungl* (sun, moon and the stars). Finally in Skáldskaparmál the word *tungl* to refer to the moon is used. See Larrington, *Edda*.

relationship to the myth outlined in this paper is indicated in the following tables.

Óðinn and Billings Mær

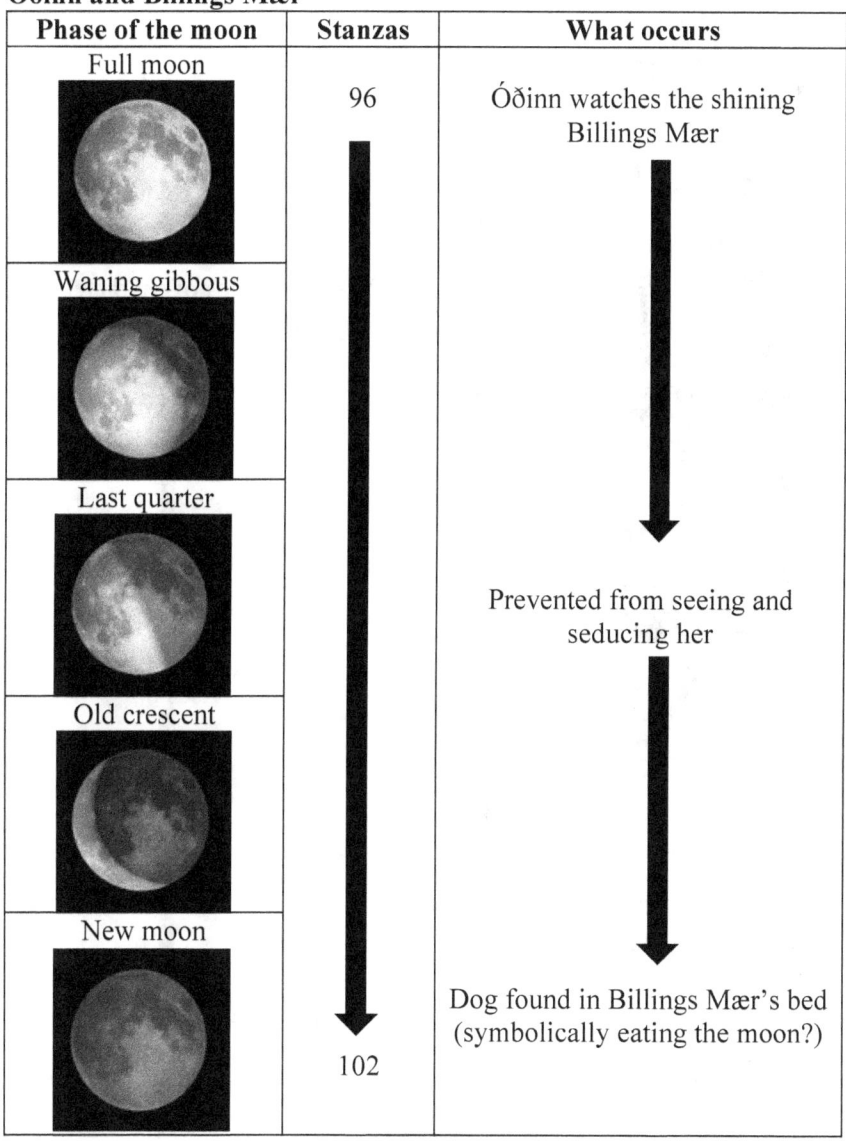

60 A Reinvestigation Into Astronomical Motifs in Eddic Poetry

Óðinn and Gunnlöð

Phase of the moon	Stanzas	What occurs
New moon	104	Óðinn single
New crescent		Personal names of characters associated with growth
First quarter		
Waxing gibbous		Óðrerir (symbolic of the moon) rises up
Full moon	110	Óðinn seduces Gunnlöð
		Óðinn breaks a ring oath, sets narrative up for repetition

On a final, if speculative note, we know that the Germanic Tribes as well as the Norse lived according to the lunar calendar.[86] This numbered 28 days.[87] This number is possibly implied in the mythic stanzas of *Hávamál*. Each of Óðinn's two adventures with the giantesses comprise of 7 stanzas, 14 when doubled to include both.[88] It has been suggested in personal communication with Gísli Sigurðsson that every half stanza may cover a full day. This equates to 28 days if each of these mythic stanzas is taken into account, which provides a perfect fit for the lunar month and therefore the movement of the moon in its full cycle. Therefore the myth of Óðinn and the two giantesses may very literally be describing the lunar cycle.

Conclusion

In conclusion, it seems likely that stargazing would have been a basic fact of life in medieval Iceland and would no doubt have informed that culture's myths and poetry as we find them written in the *Codex Regius*. As I pointed out in my introduction, such celestial myths are found across the world and are becoming increasingly studied in recent scholarship. As far as the Old Norse world and its literature are concerned, those sources in which such celestial observations have traditionally been perceived to appear are in a very fragmented state. This is in broad contrast to medieval literatures from elsewhere around the globe. However, depending on how we choose to read certain myths in Eddic poetry and *Snorra-Edda*, snippets of information and fleeting references to the sky can be uncovered. In the poem *Hávamál*, the structure of the myth of stanzas 84 and 91–104 seemingly presents itself as a duality between two complementary love stories, with a context that connects Óðinn's suffering and success in love with, respectively, the daughters of Billingr (Billings Mær) and Suttungr (Gunnlöð) in accordance with the waning and the waxing of the moon. In light of the folk belief that dictates that weddings and happiness in love are most propitious at the time of an increasing moon, it seems possible to read the relevant stanzas of *Hávamál* in view of lunar imagery. Although it

---

86 See Tacitus, *Germania*.

87 The moon's cycle from the new moon to full moon followed by its waning again to the next new moon can be expressed through the 28 days or separate phases of the moon.

88 Sigurðsson, *Eddukvæði*, p. 22 suggests that number symbolism is important elsewhere in the poem.

seems likely there was more than one tradition surrounding beliefs in the moon, we can understand the kernel of this myth as a description of what could be seen in the night sky, although the story is couched in the terms of an aphoristic tale, at the centre of which men and women's deceitfulness towards each other is portrayed with a comic twist. Hence, this mythic narrative is rooted in archaic Norse sky lore. Summing up this idea Gísli Sigurðsson posits that

> we can read the myths as unaffected descriptions of what can be seen with the naked eye during a cloudless winter night....Since we do not see but stars and an impenetrable white belt in the sky, we can let our imagination fly and change them into deities and homes of supernatural beings, good or bad ones, as we go further and lower in the sky.[89]

Therefore, although it may at first seem unusual to locate Óðinn's attempted seductions in the real-world observations of the night sky, we can choose to read the myths as a way of describing what is going on in the sky using mythological language. Thus the relevant stanzas of *Hávamál* can be treated as an astral myth centred around observations of the moon's orbit and the effect of its waxing and waning.[90]

---

[89] Sigurðsson, '*Goðsögur Snorra-Eddu*', p. 859: 'Sögunar má lesa sem blátt áfram lýsingu á því sem fyrir augu ber á björtu vetrarkvöldi...Þar sem við sjáum ekki nema stjörnur og óræett hvítt belti yfir himinum má hleypa ímyndunaraflinu á flug og breyta því í goðmögn og bústaði vætta, góðra jafnt sem illra eftir því sem fjær og neðar dregur' (my translation).

[90] This is not to claim that observations of the moon are the only factor in the development of the motif at the expense of other influences. Instead, lunar observation goes part of the way in explaining the meaning of the mythical content of the narrative.

'I specially note their Astronomie, philosophie, and other parts of profound or cunning art': The Use of Cosmic Registers by Chaucer and Others

Karen Smyth

Abstract: This paper explores the primary role that medieval authors play in promoting the language of astronomy in vernacular contexts through rhetorical and poetic translations. Focus is on Geoffrey Chaucer's 'Treatise on the Astrolabe' and how his authorising strategies compare with Adelard of Bath's earlier translation from Arabic to Latin. Notable features of literary uses of technical registers in Chaucer's other works are also commented on, as is consideration made of a wide range of texts from the thirteenth to the fifteenth centuries. What becomes evident is that there is a keen artistic interest in the Middle Ages in not simply using astronomical or astrological expressions of time but of actually understanding the significance of these registers.

Literary authors take a great interest in the language of astronomical time reckoning in the medieval period: technical registers are used for literary purposes and literary imaginings promote understanding of technical concepts. Begona Crespo Garcia observes that 'the re-emergence of the vernacular was accompanied by the functional specialisation of language according to the different areas of knowledge that were springing up under the cultural trends of the period'.[1] It was not eminent astrologers who felt the need to devise a word in the vernacular to describe the art of time reckoning. According to the *Middle English Dictionary*, the fifteenth-century monk and court poet John Lydgate first introduced the word 'computacioun' to the English vernacular in his pseudo-historical *Troy Book* (at line 2774). In the *Middle English Dictionary* there are ten

---

1 Begona Crespo Garcia, 'The Scientific Register in the History of English: a Corpus-Based Study', *Studia Neophilologica*, Vol. 76, (2004): p. 125.

Karen Smyth, '"I specially note their Astronomie, philosophie, and other parts of profound or cunning art": The Use of Cosmic Registers by Chaucer and Others', *Culture And Cosmos*, Vol. 17, no. 1, Spring/Summer, 2013, pp. 63–72.
www.CultureAndCosmos.org

subsequent uses of the word listed; used by Osborn Bokenham, John Capgrave, Ranulf Higden and further usages by Lydgate. All of these authors are vernacular writers with an interest in historical writing, who employ the term 'computacioun' across lunar, ecclesiastical, regnal and age-scheme reckonings.[2] Many writers of the late Middle Ages were interested in horological terminology: John Gower in the technical use of the unit of a 'minut'; Higden in the term 'meridian'; and Capgrave in the word 'annotacion', meaning 'notation or reckoning of time', with which he is credited for introducing in his preface to *Abbreuiacion of Chronicles*.[3] While still a geometric theory, the smallest divisions of time are employed by Lydgate in his calculation of the fate of the tragic figure of Edippus:

> The Root ytaken at the ascendent
> Trewly sought out be mynut *and* degree
> The silfe houre of his natyvyte,
> Not forgete the heuenly mansiou*n*s,
> First by Secoundes tiers a*nd* eke quartes.[4]

Literary imaginings of time therefore include technical as well as creative expressions: 'for ther is so gret diversite / in Englissh and in writyng of oure tonge'.[5]

Throughout his fourteenth-century literary texts, Geoffrey Chaucer displays this keen artistic interest not simply in using expressions of time but of actually understanding the significance of time measurements (such as the sky map of Troilus's upward ascent into the heavenly spheres that

---

[2] Hans Kurath and Sherman M. Kuhn, gen. eds., *Middle English Dictionary* (Ann Arbor: University of Michigan Press, 1956–2001), p. 478. Derek Pearsall believes that Lydgate deserves special attention for his introduction of new words into English, for 'what is more important is not that he used words first but that he used new and rare words over and over again, embedding them forever in the language'. 'Lydgate as Innovator', *Modern Language Quarterly*, Vol. 53 (1992): p. 7.

[3] *Middle English Dictionary*, p. 514 (Gower); p. 343 (Higden); p. 289 (Capgrave).

[4] John Lydgate, *The Siege of Thebes*, Part I, lines 370–74.

[5] Geoffrey Chaucer, *Troilus and Criseyde*, Book V, lines 1793–6.

occurs at the end of *Troilus and Criseyde*).⁶ J. D. North has demonstrated that it is possible to trace many 'set styles of thinking' concerning astronomical reckonings across all of Chaucer's poetry, but points out that such instances are not merely pedantically formulaic, for what 'emerges [is] an ingenious schemer, moved by a love of symmetry …a meticulous calculator'.⁷ This extensive knowledge of the cosmos has led some to contend for Chaucer's authorship of the *Equatorie of the Planetis*, a translation of a Latin work derived from an Arab text, which describes how to build and use a planetary equatorium to calculate planetary orbs and positions; but what is certain is that he composed the *Treatise on the Astrolabe*.⁸ In this *Treatise*, Chaucer's interest in and use of horological terminology suggests that this text is much more than a technical guide; and it has often been overlooked just how it demonstrates an awareness of the aesthetic and cultural significance of having command of such a technical register of expression. The majority of critical studies to date have been interested in Chaucer's technical competencies, on the nature of fourteenth-century 'science', and in the potential readership of his work. One study, however, stands out, and aligns with my interests in Chaucer's literary voice in his *Treatise*, and that is the way in which Jenna Mead explores 'the cultural valency of astrology', producing a more 'sophisticated analysis of the vernacular context of Chaucer's text'.⁹

Chaucer's claim in the *Treatise* to be merely a humble compiler, as is the case with many of the narrative guises that Chaucer adopts in his work, should be treated with some caution. Could he perhaps be engaged in a

---

6 For a detailed discussion of this sky map see Karen Elaine Smyth, 'Reassessing Chaucer's Cosmological Discourse in *Troilus and Criseyde*', *Fifteenth-century Studies*, Vol. 32, (2007): pp. 150–63.

7 J. D. North, *Chaucer's Universe*, 2nd ed. (Oxford: Oxford University Press, 1990), p. ix.

8 *The Equatorie of the Planetis* was discovered in 1952, and it shares very similar language and continues many of the ideas that are in *The Treatise of the Astrolabe*. For a discussion of the arguments for and against Chaucer's authorship, see Jennifer Arch, 'A Case Against Chaucer's Authorship of the Equatorie of the Planetis', *Chaucer Review*, Vol. 40, (2005): pp. 59–79. See also Kari Anne Rand Schmidt, *The Authorship of the Equatorie of the Planetis* (Cambridge: D. S. Brewer, 1993).

9 Jenna Mead, 'Geoffrey Chaucer's *Treatise on the Astrolabe*', *Literature Compass*, Vol. 3, (2006): p. 973.

broader agenda of cultural translation and the construction of literary identity? After all, Chaucer was the first to translate into a more accessible language the rules of time reckoning, just as the twelfth-century scholar Adelard of Bath established his reputation by making a similar linguistic translation (Adelard was the first to translate a treatise on the abacus from Arabic into Latin). It is worth noting that Arabic numerals can be found on astrolabes, whereas English mechanical clocks used Roman numerals, perhaps an indication of the independent development of time measurement, from the thirteenth century on, in Western Europe. Adelard of Bath also wrote about the astrolabe, but Chaucer's authorising strategies discussed below share more in common with Adelard's treatise on the abacus. That such a comparison can be made between the two is perhaps not surprising, as time reckoning is based on the principles of calculation. Chaucer, in his treatise, was the first writer to discuss in English the terms 'calcule' and 'calculer'.

While Chaucer's *Treatise* acts as an authoritative guide on how to use the astrolabe, and the meticulous calculations in his poetry testify to his competency in this field, it is the issue of literary authority, rather than mathematical proposition, which seems to be at stake when his text is examined alongside Adelard of Bath's treatise. Both authors, for example, seem keen to disclaim invention, Adelard writing in his preface, 'it was a certain great man that discovered all my ideas, not I', and Chaucer stating, 'I ne usurpe not to have founded this werk of my labour or of myn engyn. I n'am but a lewd compilator'.[10] But revealingly, Adelard qualifies his disclaimer by adding: 'the present generation has this ingrained weakness that it thinks that nothing discovered by the moderns is worthy to be received—the result of this is that if I wanted to publish anything of my own invention I should attribute it to someone else'.[11] While Chaucer's treatise draws heavily on work ascribed to Massahalla, whose propositions and conclusions can be found in numerous fourteenth-century Latin astronomical treatises, a number of recent critical studies have argued that some of the contents of Chaucer's treatise have no obvious source.[12]

---

10 Adelard of Bath, *Conversations with his Nephew: On the Same and the Different, Questions on Natural Science, and On Birds*, ed. and trans. Charles Burnett, (Cambridge: Cambridge University Press, 1998), p. 83.

11 Ibid.

Chaucer disclaims invention in his other works by citing unknown (and possibly even fictional) sources, such as in *Troilus and Criseyde*, where he gives Lollius as his authority.

Chaucer's *Treatise* contains other intriguing parallels with Adelard's text: Chaucer's is addressed to his young son, Adelard's to his young nephew; in composing their treatises Chaucer had 'condescendith to the rightfulle praiers of his friend', while Adelard had 'yielded to the request of my friends'.[13] Whether or not Chaucer knew about Adelard's treatise, the similarities between these two texts provides, in microcosm, a history of time-reckoning methods transmitted from the Arabic world to Western Europe via Latin, which was then translated into the new and unstable vernacular English, and presented either through simple rhyming and rhythmical phrases in technical texts, or through the use of colloquial time markings in literary texts.

What Chaucer's *Treatise* draws attention to is the great interest that medieval literary writers were taking in the language of time reckoning. The innovation of introducing technical terms to vernacular writings is, of course, noteworthy, but as A. C. Spearing notes with regard to Chaucer's use of these terms, it is the fusion of scientific astronomy and classical mythology transformed into an elevated poetic style that is the real achievement.[14] For example, Chaucer 'transfers the word 'orizonte' from scientific contexts (such as his own *Treatise on the Astrolabe*) to put it to poetic use: 'and whiten gan the orisonte shene' (*Troilus and Criseyde*, V. 276). Lydgate's later use of the word 'merydyen' illustrates the progression of this process of linguistic adaptation:

> whan Phebus passyd was merydyen
> And fro the south westward gan hym drawe,
> His gylte tressys to bathen in the wawe see.
> (*Siege of Thebes*, p. 174; Part III, lines 4256–8)[15]

---

12 This view was first seriously considered by Carol Lipson, '"I n'am but a Lewd Compilator": Chaucer's *Treatise on the Astrolabe* as Translation', *Neuphilologische Mitteilungen*, Vol. 84, (1983): pp. 192–200.

13 Chaucer, *Treatise*, p. 662 and Adelard, p. 91.

14 A. C. Spearing, *Medieval to Renaissance in English Poetry* (Cambridge, 1985), p. 73.

15 Ibid.

Evidence of such poetic play should not be seen as a narrative of easy linguistic or cultural translation. The principles of astronomy raised an aporia about the coexistence of Christian and pagan conceptions of the cosmos. While the astronomical clock dial presented the heavenly workings of the universe, it also invoked the zodiac. John Mortimer has noted this narrative of ambiguity in the Middle Ages and the failure to resolve it:

> If Robert Grosseteste could in the thirteenth century be violent in his hatred of astronomers, calling their teachings 'impious and profane, written at the direction of the devil', [*Hexamemeron*, V. XI.I] and Chaucer gently mocking of their methods in the 'Franklin's Tale' and the 'Wife of Bath's prologue' [*Canterbury Tales*, V, pp. 1261–96; lines 1117–35 and pp. 697–710; III, lines 609–20] it was nonetheless also possible for Boccaccio to defy the advice of Petrarch and hold a firm belief in the influences of the stars.[16]

Secular and religious coexistence is evident, even within the visual decoration of a single clock: in Wells cathedral, two knights and two Saracens on horseback above the astronomical dial *temporally* rotate, jousting with one another when the clock strikes, while below the clock is the *permanent* figure of the Risen Lord.

A promising trend has emerged in recent decades of a desire to investigate the effect of literary writers' scientific knowledge on the rhetorical tendencies in their poetry and any inherent rhetorical plays of the narrative of ambiguity contained therein.[17] Another related development in recent scholarship is the recognition of the medieval cosmos, not merely as a pseudo-scientific conception, but as a source of inspiration and innovation for a range of medieval poets (not just Boccacio), chroniclers

---

16 John Mortimer, *John Lydgate's Fall of Princes: Narrative Tragedy in its Literary and Political Contexts* (Oxford: Oxford University Press, 2005), p. 211.

17 See also, for example, O. Neugebauer, 'The Early History of the Astrolabe', *Isis*, Vol. 40, (1949): pp. 210–56; J. I. Cope, 'Chaucer, Venus and the "Seventh Sphere"', *Modern Language Notes*, Vol. 67, (1952): pp. 245–56; F. S. Scott, 'The Seventh Sphere: a Note on *Troilus and Criseyde*', *Modern Language Review*, Vol. 51, (1956): pp. 2–5; P. A. Dronke, 'The Conclusion of *Troilus and Criseyde*', *Medium Aevum*, Vol. 33, (1964): pp. 47–52; and North, *Chaucer's Universe*.

and philosophers in their imaginings about religion and the nature of humanity:

> in [the medieval cosmos] exotic intermixing of the spiritual and the physical, the rational and the transcendent, the finite and the infinite, and its successful incorporation of the best of pagan classicism together with the incarnation on Earth of the Creator God of Genesis [the medieval cosmos], constituted one of the most dynamic and far reaching developments in the history of human thought. Quite simply, without those Angels acending the spheres, so much of the content and metaphor of modern science and civilisation would not have existed.[18]

This new scholarly perspective is well illustrated by Chaucer's shifting use of the terms 'calcule' and 'calculer', which he transferred from the context of his technical astronomical treatise into poetic form by punning on the word 'calcule' in naming the character Calchas in *Troilus and Criseyde*: Calchas engages in the mathematical processes of prognostication. Even where time is perceived by more subjective means, as relative rather than empirical, the poetic impulse for context-dependent expression is still evidently used for literary effect. For example, in Chaucer's *Knight's Tale*, astrological unequal hours detail the exact time of the prayers of Palamon—'although it nere nat day by houres two' (line 2211); of Emily—'the thridde houre inequal that Palamon, / bigan to Venus temple for to gon' (line 2272); and of Arcite—'the nexte houre of Mars folwynge this' (line 2367). These hours are specified in order to time the prayers to coincide with the hour governed by Venus (in Palamon's case), Diana (in Emily's) and Mars (in Arcite's), thus also indicating the pagan deities that influence and inform the actions of these characters. Yet, in addition, there is obviously a keen literary consciousness at work in the use of specific astrological hours and the counting of hours in order to chart the activities of the characters. It enables Chaucer to give a sense of the continuity of time (and therefore action), a sense of sequence (and therefore causal connections), and a sense of quick progression (and therefore of the pace of narrative development). The result is not a digression in the narrative,

---

18 Allan Chapman, *Gods in the Sky: Astronomy, Religion and Culture from the Ancients to the Renaissance* (London: Channel 4 Books, 2002), p. 217.

which prayer, by virtue of its association with retreat from this world, can create; rather the prayers, located in a specific time, become a central part of the action.

Chaucer's eagerness to experiment with terms concerned with the cosmos is perhaps best accounted for by Arno Borst's assertion, that 'the ideal of eloquence also included the measured use of time'.[19] Borst notes that in medieval iconography the symbol of the mastery of speech, the parrot, is displayed alongside the symbol of proper measure, clocks, in the later period. In other words, there is a consciousness of the association between mastery of expressions of time measurement and ideals of eloquence. In the sixteenth century this trend is singled out as the most noteworthy feature with regard to Chaucer's and Lydgate's verse: 'I specially note their Astronomie, philosophie, and other parts of profound or cunning art…it is not sufficient for poets to be superficial humanists: but they must be exquisite artists, and curious uniuersal schollers'.[20]

It was not until the first half of the seventeenth century, however, that any kind of relationship between clock expressions of time and human experience of time became a prominent literary motif. René Descartes is usually credited as 'the first major thinker to incorporate the clock analogy into a philosophical system'.[21] From the seventeenth to the mid-eighteenth centuries the clock metaphor, where the workings of the world become analogous to the mechanisms of the clock, became a prominent concept for philosophers, theologians and poets. However, in Lydgate's *Testament*, a mechanical clock serves as a metaphor for the human frailty of 'gerysh' fickleness which: 'lyk a phane, ay turning to and fro, / or like an orloge whan the peys is goo'.[22] Meanwhile, almost a century earlier, in the tract

---

19 Arno Borst, *The Ordering of Time: from the Ancient Computus to the Modern Computer*, trans. Andrew Winnard, (Chicago: University of Chicago Press, 1994), p. 269.

20 Sixteenth-century author and scholar, Gabriel Harvey (c. 1545–1630). The quotation is dated 1585 and cited in Ann Astell, *Chaucer and the Universe of Knowing* (Ithaca: Cornell University Press, 1996), p. 1.

21 Samuel L. Macey, *Encyclopaedia of Time*, Vol. 810 of the Garland Reference Library of Social Sciences (New York: Taylor & Francis, 1994), p. 113.

22 John Lydgate, 'The Testament of Dan John Lydgate', in *Minor Poems of John Lydgate. Part I*, pp. 329–62. Lydgate's allusion, however, contains multiple technical difficulties: the weight in a clock imparts constant rather than 'gerysh'

Livre du Ciel et du Monde (1377), the ecclesiastic and mathematician Nicholas Oresme employs a metaphor of the universe as a vast mechanical clock regulated by God so that 'all the wheels move as harmoniously as possible' whatever the season or time of night or day, always at a steady pace, never stopping.[23] As John Scattergood remarks:

> it is clear that Oresme knows precisely what he is about: orderly movement depending on power and resistance are at the heart of the mechanics that account for clockwork. So God became a sort of divine clockmaker: having constructed the 'horologe' and set it in motion he could leave it to move by itself.[24]

This is irrefutable evidence that the literary imagination in the Late Middle Ages perceived a relationship between clock mechanics and human experience of time.[25]

But does the clock metaphor predate the technology? In their study of Chaucer's *Book of the Duchess*, for instance, Bolens and Taylor reveal that the thirteenth-century cathedral architect, Villard de Honnecourt, not only designed clock works but also drew relationships between them and the Wheel of Fortune, which is so influenced by the astrological gods. F. C. Haber, in an article on the cosmological clock metaphor, put forward an interesting thesis that decorations on European astronomical clocks were

---

movement, while the weights if they were to 'goo' would result in the clock stopping. These inaccuracies imply a not yet fully realised comprehension of the technology.

23 Lynn Thorndike, *History of Magic and Experimental Science*, Vol. 3 (New York: Macmillan, 1934), p. 405.

24 Scattergood, 'Writing the Clock', p. 464.

25 For a discussion of such a 'polysemous signification of late medieval clocks and the way in which their makers, whatever their actual intent, succeeded in linking them firmly to human concerns', see Nancy Mason Bradbury and Carolyn P. Collette, 'Changing Times: The Mechanical Clock in Late Medieval Literature', *Chaucer Review*, Vol. 43, (2009): pp. 351–75. Bradbury and Collette focus particularly on the complex series of clock metaphors in Froissart's *L'Orloge Amoureus* (1368) and Henry Suso's *Horlogium Sapientiae* (mid-fourteenth century) and the Middle English early fifteenth-century version, *Orlogium Sapientiae or The Seven Poyntes of Trewe Wisdom*.

intended as metaphors for the workings of the universe and suggested that it was these medieval visual symbols which led to the mechanistic world metaphor in seventeenth-century literature. Stephen Nichols discusses the relationship between the mathematical divisions of propositions on the astrolabe and the teleological worldview that the astrolabe was meant to confirm.[26] Awareness of the metaphorical relationship between human experience of time and technical machinations of time measurement existed before the clock analogy became a formulaic motif, due in no small part to medieval literary imaginings of the cosmos.

---

26 Guillemette Bolens and Paul Beekman Taylor, 'Chess, Clocks, and Counsellors in Chaucer's Book of the Duchess', *Chaucer Review*, Vol. 35 (2001): p. 284; F. C. Haber, 'The Cathedral Clock and the Cosmological Clock Metaphor', in J. T. Fraser and N. Lawrence, eds., *The Study of Time II* (New York: Springer-Verlag, 1975), pp. 399–416; Stephen Nichols, 'The New Medievalism: Tradition and Discontinuity in Medieval Culture', in Marina S. Brownlee et al., eds., *The New Medievalism* (Baltimore: John Hopkins University Press, 1991), pp. 1–28.

Spellbound: The Astrological Imagination of Washington Irving

Kirk Little

Abstract: Washington Irving's interest in the preternatural is well known from his gothic tales of colonial New York, notably *The Legend of Sleepy Hollow*. The lesser known *Legend of the Arabian Astrologer* tells the story of an encounter between a ninth-century Moorish King and his mysterious astrologer, conflating astrology and magic. Written during the early nineteenth–century revival of astrology in England, Irving characterizes the astrologer as magician, a notion which took on increased importance later in that century with the French occult revival of the 1850s and the 'Theosophical Enlightenment' of Madame Blavatsky. His tale is placed in the context of his literary career and astrology's cultural status at the time. Irving's limited knowledge of judicial astrology and its history led him to focus on its expression as a form of Egyptian magic. Irving paints a fairly accurate picture of the use of astral magic, including talismans, a major expression of the Arabic transmission of astrology to the medieval West. As such, his tale provides a vital link to the revival of magic and astrology in the nineteenth century as practiced by the Order of the Golden Dawn and mages like Aleister Crowley. This is the hard end of the Talismanic Magic and Astrology link that Irving pre-echoes in his story.

Rip Van Winkle and Ichabod Crane, Washington Irving's two most indelible characters, inhabit a unique space in the American literary imagination. Rip was a time traveler long before science fiction writers made such flights a commonplace idea, while Ichabod's near escape from the Headless Horseman was a cinematic page turner decades before movies were invented. The story has lent itself to numerous film adaptations, most recently as a horror film directed by Tim Burton and starring Johnny Depp, who is a police inspector investigating a number of murders in the village of Sleepy Hollow. Whether as a film or short story, spooky, supernatural elements are part of its enduring appeal. Occult forces hold both men spellbound, passive spectators to their own fate. They inhabit liminal worlds, fictional landscapes, which, not unlike Thomas Hardy's Wessex,

---

Kirk Little, 'Spellbound: The Astrological Imagination of Washington Irving', *Culture And Cosmos*, Vol. 17, no. 1, Spring/Summer, 2013, pp. 73–92.
www.CultureAndCosmos.org

would have been simultaneously familiar and strange to its readers. Familiar because the Hudson River valley was as commercially and geographically significant to early Americans as the Mississippi River would be to later generations; strange, because pockets of the region remained a cultural holdover of the early Dutch inhabitants. Irving's fictional Catskill villages are populated with quaint characters, whose customs, language and folkloric beliefs provided unsettling experiences for a new nation of people who still questioned what it meant to be an American. After an evening of gamboling with Dutch elves, Rip woke up to a transformed country. In Irving's allusive prose, woodland elves and spectral presences serve as explanatory metaphors for the ravages of time. Perhaps not so strangely, Irving wrote these stories while living in England.

Irving intuitively understood how his eerie tales struck a cultural nerve with a public not quite recovered from the Salem witch trials of 1692. In *The Legend of Sleepy Hollow*, we learn that school master Crane counted among his few possessions a copy of Cotton Mather's *History of New England Witchcraft*, 'in which, by the way, he most firmly and potently believed',[1] along with 'a *New England Almanac*, and a book of dreams and fortunetelling'.[2] Irving's fascination with the preternatural traveled with him years later to Spain, where he wrote a forgettable biography of Columbus, but once again hit mythic pay dirt with *The Alhambra: A Series of Tales and Sketches of the Moors and Spaniards*.[3] Sifting the Spanish archives may have provided him with historical context, but gossiping with the locals and listening to their improbable yarns enabled him to create the atmospherics so necessary to Irving's best stories.[4] Among the tales, *The Legend of the Arabian Astrologer* permits us to glimpse how one of America's best-known early nineteenth century authors came to grips with astrology's mysteries and provides one of the keys to its enduring appeal.

---

1 Washington Irving, *The Complete Tales of Washington Irving*, ed. Charles Neider, (New York: Da Capo Press, 1975), p. 36.

2 Irving, *Tales*, p. 54.

3 Washington Irving, *The Alhambra: A Series of Tales and Sketches of the Moors and Spaniards* (New York: John Alden, 1881).

4 Van Wyck Brooks, *The World of Washington Irving* (New York: E. P. Dutton & Co., 1944), pp. 252–53.

This tale from the Alhambra deserves our attention because Irving's depiction of the astrologer as magician took on increased significance during the nineteenth century. Magic rituals, as part of Masonic lore were transmitted to the American colonies during the middle third of the eighteenth century, under the authority of the London grand lodge.[5] By Irving's day (indeed, his brother William was a Lodge master in New York City, who 'helped his non-Masonic brother write *Salmagundi*'.[6]) Freemasons 'still found the symbols and structures of learned magic valuable', even if they 'never explicitly claimed any kind of occult powers for their secrets'.[7] Others in Europe and America took their magical pursuits more seriously. As that century progressed, there was a 'group of thinkers…who, inheriting the mantle of eighteenth-century Illuminism, had begun to formulate a philosophical context for modern astrology as magic'.[8] Eliphas Levi was part of that group; his magical and Kabbalistic theories led to a French occult revival in the 1850s. Magic was in the air. A shadow culture of spirit-seers, table rappers, phrenologists, and palmists plied their trades in many large cities, a virtual haven for those souls alienated by the century's increasing mechanization and materialism.[9] During the last quarter of that century, Madame Blavatsky and her acolytes produced the 'Theosophical Enlightenment', which in turn helped produce the modern revival of astrology.[10] Eventually, Magic and Astrology were to become especially prominent in the late nineteenth century occult groups like the Order of the Golden Dawn and in the practice of mages like Aleister Crowley. Thus, Irving's Arabian tale may be seen in retrospect as

---

5 Steven C. Bullock, *Revolutionary Brotherhood: Freemasonry and the Transformation of the American Social Order, 1730–1840* (Chapel Hill, NC: University of North Carolina Press, 1996), Ch. 2.

6 Ibid., p. 153.

7 Ibid., p. 20.

8 Nicholas Campion, *A History of Western Astrology, Volume II The Medieval and Modern Worlds* (London: Continuum Books, 2009), p. 222

9 Eugene Taylor, *Shadow Culture: Psychology and Spirituality in America* (Washington DC: Counterpoint, 1999) Ch. 7.

10 Patrick Curry, *A Confusion of Prophets: Victorian and Edwardian Astrology,* (Collins & Brown, London, 1992), Ch. 5; Campion, *History*, Ch. 16.

a harbinger of the West's increasing preoccupation with the mysteries of the East. His story of the Arabian astrologer, like a swaying shadow lantern, casts an odd but interesting light on astrology's recent history.

Irving's Career
Born in 1783, the year Americans achieved their independence from England, Irving spent much of his literary career smoothing the feathers ruffled by his native countrymen's audacious move. Named for the American Revolution's indispensable leader,[11] he greatly admired England and at various times was accused of harboring Tory sentiments. Yet his literary style embraced the American vernacular and embodied a painterly preoccupation with the untamed natural environment. At the time, his fellow Americans were more preoccupied with buying and selling land to each other and according to one journalist, demonstrated an 'almost universal ambition to get forward'.[12] Such eager commercialism was hardly conducive to producing a national literature.

For his part, Irving didn't quite fit the grasping mold. After his first few books failed to support him, he moved to England in 1815 to escape his family's pressure to join their business. Writing almost in desperation to his friend Henry Brevoort back in the States, he acknowledged as much: 'My mode of life has unfortunately been such as to render me unfit for almost any useful purpose. I have not the kind of knowledge or the habits that are necessary for business or regular official duty. My acquirements, tastes and habits are just such as to adapt me for the kind of literary exertions I contemplate'.[13] A wide reader, he knew the sorry state of American literature.

Much of the literature produced by Americans during their early national period, roughly the forty years following the formation of the United States, 'were designed for untutored readers....Their plots were straightforward, their vocabulary undemanding, and their syntax

---

11 Ron Chernow, *Washington: A Life* (New York: The Penguin Press, 2010), p. 584.

12 *Niles Weekly Register*, 9 (1815), p. 238, quoted from Gordon S. Wood, *Empire of Liberty: A History of the Early Republic, 1789-1815* (New York: Oxford University Press, 2009), p. 3.

13 *Letters of Washington Irving to Henry Brevoort*, Vol. II, ed. George S. Hellman, (New York: G. P. Putnam's Sons, 1918), 10 July 1819, p. 309.

unsophisticated'.[14] More discerning readers continued to favor English novels and plays over the output of their countrymen. Americans, fiercely independent in their politics, remained cultural colonists of Britain well into the nineteenth century. National pride searched in vain for someone who could match the enticing legends of Walter Scott, by far the most popular writer during the first quarter of the nineteenth century. His books sold in the thousands in the United States and by 1823 totaled a quarter million. Scott's popularity was such that no fewer than thirty five towns in the United States were named Waverley, in honor of his most well known novel. While a few thought his 'diseased and perverted taste for the luxurious and aristocratic' had no place in a fledgling democracy, many more, especially in the south, thrilled to his gothic romances and named their daughters Rowena and their side-wheel steamers *Rob Roy*.[15]

In 1817, Scott, secure in his fame, invited a visiting American writer to his home in Abbotsford, Scotland. He had been impressed by the young author's *Knickerbockers' History of New York*, and the two men talked for days about Scottish myths and lore. As it turned out, his visitor Washington Irving was at odds with himself; his mother had just died a few months earlier and the family business was about to go bankrupt. Unknown to either man, he was also about to rescue American literature from complete domination by English literary conventions. In 1820, the English critic Sidney Smith sneered 'Who reads an American book?'[16] As if in response, Irving produced a trans-Atlantic best seller and rapidly achieved literary fame with his *The Sketch Book of Geoffrey Crayon, Gent.*,[17] which was published serially throughout 1819 and 1820.

The book's success launched his literary career. Mixing fanciful stories from his childhood in New York with thumbnail sketches of English places, social customs and singular characters he met on his travels, it was received with critical acclaim on both sides of the Atlantic. While its travelogue studies continue to exude a certain charm, its enduring fame rests on two stories: *Rip Van Winkle* and *The Legend of Sleepy Hollow*.

---

14 Wood, *Empire*, p. 570.

15 Russell Blaine Nye, *The Cultural Life of the New Nation 1776–1830* (New York: Harper & Row, 1960), pp. 254–55.

16 Ibid., p. 257.

17 Washington Irving, *The Sketch Book of Geoffrey Crayon, Gent.* (New York: The Heritage Press, 1939).

Like much of his best writing, these tales retain a power to move readers through Irving's ability to evoke not just a place, but an atmosphere. Readers, not unlike Rip and Ichabod Crane, are lightly transported, almost dreamlike to places half familiar, yet strange. As Irving explained to Brevoort:

> I fancy much of what I value myself upon in former writing, escapes the observation of the great mass of my readers, who are more intent upon a story than the way in which it is told. For my part, I consider a story merely as a frame on which to stretch my materials. It is the play of thought, and sentiment, and language; the weaving in of characters, lightly, yet expressively delineated...[18]

It should not surprise us to learn that Irving flirted with the idea of becoming a painter and counted several artists among his friends. His evocative gifts, his sentimentality, and his poignant nostalgia have led literary scholars to label him a romantic, but beneath the dreamy surface lay a canny pragmatist whose stories frequently involve a comeuppance for the weak and credulous. If his English readers admire his grace, charm and wit, his American readers equally enjoy Irving's wry realism.

These gifts are on display, though in somewhat diminished form, in his *The Legend of the Arabian Astrologer*. Published in 1832 upon Irving's return to America after a seventeen-year absence, the story had been written in 1829 after several months living in the Moorish ruins of the Alhambra. Throughout his childhood, Irving had been enchanted by stories of Spain and the Moors; his travels there in 1826 followed another trough in his career and he hoped to find inspiration amongst the crumbling magnificence of this former Islamic fortress. As Irving languidly informed Brevoort, 'I lounge with my book about these oriental apartments or stroll about the courts and gardens and arcades, by day or night with no one to interrupt me. It absolutely appears to me like a dream; or as I am spell bound in some fairy palace'.[19] As someone who took inspiration from his surroundings, we must remember Irving's Alhambra was quite different from its current gleaming incarnation. According to Van Wyck Brooks, by the late 1820s, the palace

---

18 Hellman, *Irving Letters*, 11 December 1824, pp. 398–99.

19 Ibid., 23 May 1829, p. 426.

had not yet undergone any restoration, and the roses and weeds grew wild on the terraces and gates, while the fires of beggars smoked the Moorish arches and criminals hid in the grottoes and holes in the walls. A ragged brood of peasants and invalid soldiers inhabited the corridors and courts...while gypsies strayed in from the caverns and hills...[20]

While there among the ruins, Irving was regaled by the locals with tales of talismans, charms, and magic spells.[21] These elements figure heavily in his tale of the Arabian astrologer; the technical language of judicial astrology does not. To see why, we need to briefly examine the cultural status of astrology in the United States and England during Irving's formative years.

Early Nineteenth Century Astrology
At the close of America's revolutionary era in the late 1780s, when Washington Irving was a little boy, Ezra Stiles, patriot, scholar, clergyman and the President of Yale, expressed puzzled dismay at the persistence of astrological and occult beliefs among some of his newly liberated countrymen. How, in the wake of the Enlightenment's greatest political triumph of reason and natural law, could men still practice magic? Fortunately, according to Stiles, 'In general, the System is broken up, the Vessel of Sorcery shipwreckt, and only some shattered planks and pieces disjoyned floating and scattered on the Ocean of...human Activity and Bustle'.[22] Of course, Stiles' watery metaphor expressed an obvious truth: the world view which had produced astrology's mid-seventeenth-century flowering had largely sunk beneath the surface of polite and learned discourse. Its successor, Newton's clockwork universe, had little use for signs and divination, but many people still did.

As Patrick Curry has reminded us, astrology did not disappear during the eighteenth century, however, its practice 'was confined almost entirely to the semi-literate labouring class, in the form of popular beliefs

---

20 Brooks, *World*, p. 252.

21 Irving, *Alhambra*, pp. 86, 158; Brooks, *World*, p. 253.

22 Jon Butler, *Awash in a Sea of Faith: Christianizing the American People*, (Cambridge, MA: Harvard University Press, 1990), p. 88.

concerning the phases of the Moon and other readily visible phenomena'.[23] Judicial astrology, the practice of reading horoscopes, still claimed a few practitioners, but in the late eighteenth century, the only form of astrology most Americans encountered was to be found in farmer's almanacs. Naturally, these journals, like their seventeenth-century predecessors contained monthly planetary ephemerides and even the 'Zodiac Man', the splay-footed melothesic figure that indicated which parts of the body were ruled by which astrological signs. Still, the popularity of these publications was not due to any occult yearnings of the yeomanry, rather to their perceived practicality; in addition to medical remedies and homiletic advice, almanacs assisted a nation of farmers with a timetable for planting and sowing their crops.[24] Metaphysics took a back seat in the bustling American caravan of commerce.

Glimmers of astrology's eventual late Victorian revival could be detected in London, if one knew where to look. Apparently, Irving did, since he informed a correspondent the inhabitants 'still believe in dreams and fortune telling, and an old woman who lives in Bull-and-Mouth Street makes a tolerable subsistence by detecting stolen goods, and promising girls good husbands. They are apt to be rendered uncomfortable by comets and eclipses...'[25] A growing number of enthusiasts read short lived journals such as John Cornfield's *The Urania, or, the Philosophical Magazine*, in 1814, or a decade later Robert Cross Smith's *The Straggling Astrologer*, which had the audacity to publish the King's horoscope and chide his Queen for infidelity. In one historian's words, '*The Straggling*

---

23 Curry, *Confusion*, p. 10.

24 Astrology in eighteenth-century North America: Herbert Levanthal, *In the Shadow of the Enlightenment: Occultism and Renaissance Science in Eighteenth Century America* (New York: New York University Press, 1976); William D. Stahlman, 'Astrology in Colonial America: An Extended Inquiry', *William and Mary Quarterly*, Volume 13, (October 1956): pp. 551–63; Jon Butler 'Magic, Astrology, and the Early American Religious Heritage, 1600-1760', *American Historical Review*, Vol. 84, (April 1979); and 'The Dark Ages of American Occultism, 1760–1848', *The Occult in America: New Historical Perspectives,* ed. Howard Kerr and Charles L. Crow, (Urbana, IL: University of Illinois Press, 1983); Jon Butler, *Awash in a Sea of Faith* (Cambridge, MA: Harvard University Press, 1992). Almanacs: Robb Sagendorph, *America and her Almanacs: Wit, Wisdom & Weather 1639–1970* (Boston: Little, Brown and Company, 1970).

25 Curry, *Confusion*, p. 9

Astrologer exploited the combination of high moral tone, sexual titillation, and snobbery that has never failed the popular press'.[26] No doubt such impertinence contributed to the passage in Britain of The Vagrancy Act of 1824; section 4 of that statute specifies that it 'applies to "every Person professing to tell fortunes or use any subtle Craft, Means or Device, by Palmistry or otherwise, to deceive or impose upon His Majesty's Subjects."'[27] Apart from eccentrics like Cornfield and Smith, there were artists like John Varley, William Blake's friend and a keen student of astrology, as well as more scientifically minded practitioners such as James Wilson, who compiled his *Dictionary of Astrology* published in 1819.[28]

Other writers presented astrology as part of a spectrum of occult studies; perhaps the most notable was Francis Barrett, who published his *The Magus, or Celestial Intelligencer* in 1801. As Joscelyn Godwin notes, Barrett produced a 'fairly complete occult manual, which taught the principles of arithmology and correspondences, planetary and Kabbalistic magic, and scrying technique...'[29] We do not know whether Irving came across any of these publications, but at the very time he was wandering amidst the decaying exotica of the Alhambra, his literary hero (and now friend) Walter Scott seems to have taken inspiration from such sources. He had just published the second of his Waverley novels, *Guy Mannering*, which was originally conceived along astrological lines. Wisely, Scott understood that astrology could 'not now retain influence over the general mind sufficient even to constitute the mainspring of a romance',[30] and he dropped the astrological theme after the fourth chapter. While Scott's

---

26 Joscelyn Godwin, *The Theosophical Enlightenment* (Albany, NY: State University of New York Press, 1994), p. 143.

27 Curry, *Confusion*, p. 13.

28 For the nineteenth century revival: Curry, *Confusion*; Godwin, *Theosophical*, Ch. 7; Ellic Howe, *Astrology: A Recent History Including the Untold Story of Its Role in World War II*, [Originally published in Great Britain as *Urania's Children*] (New York; Walker and Company, 1967), Ch. 3; Campion, *History*, Chs. 13 and 14.

29 Godwin, *Theosophical*, p. 119.

30 Walter Scott, *Guy Mannering* (London, 1829) *Introduction*, p. x, (from American facsimile of 1829 edition, published by Aldine Book Publishing Co., Boston, no date)

protagonist may have been 'bewildered amid the arithmetical labyrinth and technical jargon of astrology',[31] it is clear the author had a good grasp of judicial astrology in a way that Irving apparently did not; he understood, at least in broad outline, the tradition of Lilly and Culpeper, perhaps England's two best known astrologers from the seventeenth century.[32] Irving's ideas of astrology are hazier; they revert to astrology's roots in the magical tradition. His astrologer is more a magician than a prognostic technician; an Egyptian, not an Englishman. Let us turn to his tale.

The Arabian Astrologer
Irving's tale of an Arabian astrologer is easily told in outline. It concerns an aged Moorish King, Aben Habuz, who had ruled his kingdom of Granada for decades, but now felt weary and desirous of rest. Young rivals threatened the old king, who felt besieged by his many enemies, which kept him in 'a constant state of vigilance and alarm'.[33] Almost miraculously, an astrologer appears in the guise of 'an ancient Arabian physician'[34] named Ibrahim Ebn Abu Ayub, who arrives at the King's court and quickly solves his problems by installing a magic talisman which is able to detect approaching enemies. The King generously rewards his mysterious guide by allowing him to build a 'sumptuous subterranean palace'[35] on the grounds of his kingdom. After a short while, the astrologer asks for more earthly rewards, which the King grudgingly grants. One day, the King's soldiers, alerted to the presence of an enemy by the astrologer's talisman, return from reconnoitering the surrounding area with a 'damsel of surpassing beauty'.[36] The astrologer warns the King that she is trouble

---

31 Ibid., p.20.

32 Campion, *History*, Ch. 11; Patrick Curry, *Prophecy and Power: Astrology in Early Modern England* (Oxford: Polity Press, 1989); Benjamin Woolley, *Heal Thyself: Nicholas Culpeper and the Seventeenth-Century Struggle to Bring Medicine to the People* (New York: Harper Collins, 2004).

33 Irving, *Tales*, p. 498.

34 Ibid,.

35 Ibid., p. 503.

36 Ibid., p. 504.

and insists he be allowed to keep her for himself to 'refresh' his mind 'weary with the toils of study'.[37] He assures the King that he has counter spells to protect himself. When the King insists she stay with him instead, the two men have a falling out and trouble quickly ensues for the King. He again begs the astrologer for help. Beguiling the King with promises to build him a 'delectable palace' in an earthly paradise, the wily astrologer tricks him into an agreement that if he builds such a place, the King will grant him 'the first beast of burden, with its load, which shall enter the magic portal of the palace'.[38] The King agrees, but as they approach the gateway, the astrologer distracts the King with talk of magic talismans while the lovely young maiden, riding a palfrey, is carried across the threshold. When the astrologer claims his prize, the King becomes enraged and offers him all of his wealth. The insulted astrologer refuses, and angrily 'smote the earth with his staff, and sank with the Gothic princess through the centre of the barbican'.[39] In desperation, the King orders his men to dig for the maiden, but all attempts to retrieve his beloved princess are unsuccessful. The unhappy King soon dies. In the ensuing years, the Alhambra is built on the site of this earthly paradise.

Two themes emerge from reading his tale: Irving seems to think of astrology as primarily an Egyptian invention; and because he lacks a technical understanding of judicial astrology, he equates astrology and magic. I shall examine each of these themes.

The 'Egyptian' Astrologer
Like many of Irving's leading characters, the Arabian astrologer slips into the story almost unnoticed. He arrives like some figure lifted from one of Irving's boyhood reading pleasures, the *Arabian Nights*: 'His gray beard descended to his girdle, and he had every mark of extreme age, yet he had travelled almost the whole way from Egypt on foot, with no other aid than a staff marked with hieroglyphics'.[40] Though he is described as 'an ancient Arabian physician', he is no mere medical practitioner, but a man who had

---

37 Ibid.

38 Ibid., p. 507.

39 Ibid., p. 509.

40 Ibid., p. 498.

spent years in Egypt 'studying the dark sciences, and particularly magic, among the Egyptian priests'.[41] As he assured the King, even the source of his magic was derived from a sacred book the astrologer had seized 'with a trembling hand' from within a 'sepulchral chamber'[42] inside one of the famous pyramids. His prestige derives from his Egyptian magic, not his astrological expertise.

His Egyptian roots are equally reflected in his surroundings. Initially the humble wise man requested nothing more than an unadorned cave as his reward for helping the King turn away his enemies. In no time, Ebn Abu Ayub persuaded the King not once but twice to transform his living quarters into a 'suitable hermitage' fit for an Egyptian astrologer. First, the astrologer 'caused the cave to be enlarged so as to form a spacious and lofty hall, with a circular hole at the top, through which, as through a well, he could see the heavens and behold the stars even at mid-day. The walls of this hall were covered with Egyptian hieroglyphics with cabalistic symbols, and with the figures of the stars in their signs'.[43] We must pause to consider this, the only explicit reference to astrology in the entire tale. No specific signs or planets are mentioned, I would suggest, because apparently Irving could not convincingly deploy them to enhance the narrative charm of his tale. Instead, he reverts to his gift for evoking an exotic atmosphere.

Later, after a second magical intervention, the delighted King granted the astrologer's request to turn his 'astrological hall' into a pleasure palace 'furnished with luxurious ottomans and divans, and the walls to be hung with the richest silks of Damascus'.[44] There are more demands. Even 'the light of the sun…is too garish and violent for the eyes of an old man'; therefore, the astrologer 'caused the apartments to be hung with innumerable silver and crystal lamps, which he filled with a fragrant oil prepared according to a receipt discovered by him in the tombs of Egypt'.[45] The astrologer's penchant for luxurious things suggests his own kingly status, just as his name is almost an anagram of the King's name. He also

---

41 Ibid.

42 Ibid. p. 500.

43 Ibid. p. 499.

44 Ibid. p. 502.

45 Ibid.

understands his Egyptian 'wisdom' places him on an equal, if not superior footing to a mere king. Of course, as a modern reader can clearly see, Irving's depiction also neatly fits an early nineteenth century western predilection for viewing Oriental magi as more interested in material trappings than any genuine pursuit of wisdom.

Irving's choice of Egypt as the source of Ebn Abu Ayub's wisdom probably had more to do with recent events than any deep understanding of astrological history. Egypt had re-entered the European imagination during Napoleon's campaign of 1799, when his French troops unearthed what appeared to be a broken slab of black basalt (it turns out to be granite) while digging trenches near Rashid or Rosetta. The discovery of the Rosetta Stone, with its trilingual text of Greek, Hieroglyphic, and Demotic Egyptian scripts, promised a 'final key to Egypt's lost history'[46] if translators could crack its hieroglyphic code. While hieroglyphics had long fascinated and stymied European scholars, the accomplishment of that task was achieved by Jean Francois Champollion in 1822 after more than fourteen years of labor. The publication of Champollion's 1824 work *Précis du Systeme hieroglyphique* created a minor sensation and further piqued the public's interest in this land of ancient mysteries. In 1826, Champollion was named a conservator of the Louvre Museum's Egyptian collection, which opened in late 1827. The following year, he conducted the first systematic survey of Egypt's monuments, archeology and history, which led to his appointment as a professor of a chair in Egyptian history at the College du France created especially for him. Then too, unfolding events in the Near East reminded the wider public of this land of mysteries. A diplomatic crisis in 1831 had alarmed both the great powers of Europe and the Ottoman Sultan, when the French government backed the empire building pretensions of the Egyptian governor Mehemet Ali.[47] Such events kept Egypt in the news and stoked the desires of the public to know more about this strange culture.

The Astrologer as Magician
An astrologically informed reader searches in vain for any evidence of horoscopic astrology in Irving's tale. Instead, one finds the ancient

---

46 Lionel Casson, *Ancient Egypt* (New York: Time-Life Books, 1965), p. 16.

47 William Langer, *Political and Social Upheaval: 1832–1852* (New York: Harper & Row, 1969), pp. 285–89.

astrologer's power and prestige lie in his magical accomplishments. To solve the old King's recurring problem with his enemies, Ebn Abu Ayub informed the King that 'I am instructed in all the magic arts, and can command the assistance of genii to accomplish my plans'.[48] In no time, he had created a replica of 'the Talisman of Borsa':

> He caused a great tower to be erected upon the top of the royal palace...the tower was built of stones brought from Egypt, and taken; it is said, from one of the pyramids. In the upper part of the tower was a circular hall, with windows looking towards every point of the compass, and before each window was a table, on which was arranged, as on a chessboard, a mimic army of horse and foot, with the effigy of the potentate that ruled in that direction, carved in wood. To each of these tables there was a small lance, no bigger than a bodkin, on which were engraved certain Chaldaic characters.[49]

This magical talisman becomes the aged King's plaything. By manipulating 'a bronze figure of a Moorish horseman, fixed upon a pivot',[50] he could detect the presence of approaching enemies and quickly dispatch them with a twist of the lance. The efficacy of the astrologer's talisman almost goes without saying. Before long, the anxious King 'thrust the magic lance into some of the pigmy effigies...it was with difficulty the astrologer could stay the hand of the most pacific of monarchs, and prevent him from absolutely exterminating his foes'.[51] Such magical feats are a long way from character delineation and predictions for the future. Here we should pause and take stock of Irving's image of astrology and astrologers by examining the connection between astrology and magic.

---

48 Irving, *Tales*, p. 500.

49 Ibid.

50 Ibid., p. 501.

51 Ibid., pp. 501–2.

Magic and Astrology
During the twelfth century, Moorish scholars in Spain retranslated many of the ancient Greek astrological and magical texts from Arabic into Latin, after which they were introduced to medieval Europe. A prominent example is the *Ghayat al-Hakim*, known in the West as the *Picatrix*, which has been described as 'an extremely comprehensive treatise on sympathetic and astral magic, with particular reference to talismans'[52] and 'the key text of magical astrology until the seventeenth century'.[53]

The *Picatrix* was written in the tenth century in Arabic and translated into Spanish and Latin in the thirteenth century. According to Richard Cavendish, a modern authority on the European magical tradition, translations from Arab authors 'were equally important in the revival of a sophisticated interest in astrology'.[54] The connection between magic and astrology in the medieval Arab world takes on increased significance for our study of Irving's tale when we realize it is set in Granada during the ninth century, during the era of Masha'allah, Al Kindi and Abu Mashar, the three greatest astrologers of the Arabian empire.[55] They provided a unique philosophical rationale for an astrology which synthesized Aristotelian Greek science with Islamic cosmological doctrines and Neo-Platonic thought. The astral magic, which informs Irving's tales, was an important part of Arabic astrology.

Simply put, astral magic employs astrological knowledge and techniques to assist the magician in bringing about certain outcomes. Astral magicians 'presuppose that continual effluvia of influences pouring down onto the earth from the stars…could be canalized and used by an operator with requisite knowledge.'[56] Because magic seeks to change one's destiny, rather than merely learn it, its methods and aims are different from

---

52 Frances A. Yates, *Giordano Bruno and the Hermetic Tradition*, (Chicago: University of Chicago Press, 1964), p. 49.

53 Campion, *History*, p. 66.

54 Richard Cavendish, *A History of Magic* (Arkana/Penguin, London, 1987), p. 65.

55 *Chronology of the Astrology of the Middle East and the West by Period*, compiled by Robert Hand, (Archive for the Retrieval of Historical Astrological Texts, no date), p. 11.

56 Yates, *Bruno*, p. 45.

judicial astrology. A key component of astral magic is the use of talismans, objects marked with words, letters, or magical signs and believed to confer on the bearer supernatural powers or protection.[57] According to Cavendish, 'An astrological image or talisman could be regarded as the container of a spirit which had been drawn down into it, so that it was a portable magical power-source'.[58] The production of talismans was a complex procedure, which was supposed to be 'made under the correct astrological conditions (under the appropriate planetary hour, and so on), but it had to be manufactured in the metal or gem related to the planetary or stellar forces being evoked in the charm'.[59] Irving is not concerned with such technicalities; however, he knew talismans are carriers of great symbolic significance. Let us examine his use of them, as well as magical symbols found in the palace of the Alhambra.

Symbolic Magic

Despite his lack of a technical understanding of horoscopic astrology, Irving's use of astrological and magical symbolism is often quite apropos. This may be attributed to two factors: his interest in learning about the myths and beliefs of the local populace of Grenada, and his handling of primary and secondary sources, including books acknowledging the importance of Arabic magical texts. We also know that he spent a considerable amount of time in Spanish historical archives between 1826 and 1831, since he produced three nonfiction works during those years; two books on Columbus and the *Conquest of Grenada*, published in 1829. This latter work alludes to the intrigues of court astrologers in recounting the history of the Islamic struggle for control for this region of southern Spain.[60] Certainly, Irving appears to understand the significance of talismans in Arabic magic. In a 'Note to the Arabian Astrologer' appended to his tale, Irving reports 'Al Makkari, in his *History of the Mohammedan*

---

57 Richard Kieckhefer, *Magic in the Middle Ages* (Cambridge: Cambridge University Press, 1989), p. 77.

58 Cavendish, *Magic*, p. 67.

59 Fred Gettings, *The Arkana Dictionary of Astrology* (London: Arkana/Penguin, 1990), p. 498.

60 See Washington Irving, *The Conquest of Grenada* (Boston: Desmond Publishing, 1899), Chapters 3 and 4.

Dynasties of Spain, cites from another Arabian writer an account of a talismanic effigy somewhat similar to the one in the foregoing legend.'[61]

Returning to our tale, we see that the replica of the Talisman of Borsa, which Ebn Abu Ayub constructed for Aben Habuz, was based upon 'a great marvel devised by a pagan priestess of old...(it) was a figure of a ram, and above it a figure of a cock, both of molten brass, and turning upon a pivot'.[62] The King understood its symbolism immediately: 'what a treasure would be such a ram to keep an eye upon these mountains around me; and then such a cock, to crow in time of danger!'[63] Thus, Irving has brought together two animals with relevant symbolism. The ram, a potent symbol of Aries, is noted for being impetuous, virile, hot-headed and headstrong; when it meets an obstacle, the ram will charge it. Mars, Aries' ruler, is the god of war. The cock is venerated in Islam, since 'it was the giant bird seen by Muhammad in the First Heaven crowing'.[64] A powerful symbol across many cultures, the cock was 'known to European antiquity variously as an animal of the sun, crowing to announce daybreak and drive off nocturnal demons, or (especially the black rooster) as a magical and sacrificial animal for subterranean powers'.[65] By announcing the arrival of daybreak, the cock or rooster also takes on an attribute of Mercury-Hermes; one mid seventeenth century image depicts the cock prancing at the feet of winged Mercury clutching his caduceus.[66] This ram and cock template for the astrologer's talisman for the King is a Mars/Mercury symbol suggesting their angry verbal exchanges, as well as a device which communicates the King's intention to make war on his rivals.

We see a similar handling of symbolism in Irving's use of the Alhambra's famous gateway. In contriving to build his King an earthly paradise, Ebn Abu Ayub 'caused a great gateway or barbican to be

---

61 Ibid., p. 511.

62 Irving, *Tales*, p. 499.

63 Ibid., p. 499.

64 Jack Tresidder, ed., *The Complete Dictionary of Symbols* (San Francisco, Chronicle Books, 2004), p. 115.

65 Hans Biedermann, *Dictionary of Symbolism: Cultural Icons & the Meanings Behind Them* (New York: Penguin Books, 1989), p. 288.

66 Ibid., p. 288.

erected…on the keystone of the portal the astrologer, with his own hand, wrought the figure of a huge key; and on the keystone of the outer arch…he carved a giant hand.'[67] The significance of these symbols is made clear in the second chapter of the collection, entitled 'The Interior of the Alhambra': 'Those who pretend to some knowledge of Mohometan symbols affirm that the hand is the emblem of doctrine, and the key, of faith.'[68] These symbols of Islam also appear to have magical import. Irving reports that 'According to Mateo, it was a tradition handed down from the oldest inhabitants…that the hand and the key were magical devices on which the fate of the Alhambra depended.'[69] Irving seems to be intimating some greater mystery to which his story points. That mystery concerns the relationship of doctrine and faith, of magic and wisdom.

Irving does not discuss that mystery; this is an Arabian tale, not a philosophical treatise. Still, his resolution of the story suggests he appreciated it. He understood that using magic to achieve base goals does not equate with wisdom. Throughout the tale Irving deftly shows how the intertwining fate of each man was dictated by their uncontrolled passions. Magic amplified those passions, but did not alter them. The astrologer's magic talisman was used in the service of the King's fear and anger, just as the King's earthly rewards brought out the astrologer's indolence and greed. Thus, each man was spellbound by the other. Ultimately, both the King and his astrologer sealed their respective fates, by their thralldom to the beautiful 'Gothic princess'. Once the astrologer disappeared beneath the ground with his beautiful prize, the King lost his magic protection; he was besieged by his enemies and soon died. For all his magic powers, the astrologer was enfeebled by his triumph over the King. Irving leaves us with a wistful image of the old astrologer 'nodding on his divan, lulled by the silver lyre of the princess.'[70] He seems more under her spell, than the reverse; earthly lust has sapped his powers.

---

67 Irving, *Tales*, p. 508.

68 Irving, *Alhambra*, p. 33.

69 Ibid., p. 34.

70 Irving, *Tales*, p. 510.

Kirk Little 91

Magic's Legacy
From this dispiriting story of lust, greed and anger Irving derives the very palace where he conceived and wrote his tale. On the site of 'this dispute between two graybeards for the possession of youth and beauty',[71] where the astrologer had once promised to build the King a 'sumptuous palace with a garden'[72] arose the Alhambra. More than simply a palace or a fortress, the Alhambra embodies the spiritual vigor of early Islam; its surroundings 'in some measure realizes the fabled delights of the garden of Irem'[73] recorded in the Koran. It is the fruit of their protracted struggle. Conceived in magic from the spectral vision conjured out of thin air by the astrologer; born in duplicity and deceit as a bedazzling vision which enabled the astrologer to steal the King's princess, the earthly Alhambra retains its most striking feature. As Irving assures his readers, 'The spellbound gateway still exists entire, protected no doubt by the mystic hand and key, and now forms the Gateway of Justice, the grand entrance to the fortress'.[74] Inside the palace, the slim, arcaded rooms shimmer with light and retain their power to transfix visitors. That enduring power, Irving suggests, may still emanate from under that gateway, where the astrologer remains in his subterranean hall lulled by his princess.

Irving has performed his own magic. In his skillful prose, 'the weaving in of characters, lightly, yet expressively delineated', he reminds us, as he did with his tales of Rip and Ichabod, of the importance of unseen forces in man, in nature and in the universe. However we choose to understand those forces, as science or magic, Irving reminds us, they contribute to his two character's divergent fates. His nineteenth-century successors would have it both ways; astrology would remain a species of magic for some and a misunderstood science for others. Irving was unconcerned with such things; he was above all, an imaginative writer and his singular tale remains a vital reminder of the persistence of astrological imagination. For historians of astrology, however, Irving and his *Legend of the Arabian Astrologer* serve as a kind of barometer for his contemporary understanding of astrology and magic. Tellingly, he leaves us with a

---

71 Ibid., p. 509.

72 Ibid., p. 506.

73 Ibid., p. 510.

74 Ibid.

drowsy image of 'the astrologer, bound up in magic slumber by the princess',[75] as if to suggest that astrology, like Rip Van Winkle would need to be nudged awake to confront a very different world, than the one it fondly recalled from its bygone days.

---

75 Ibid., pp. 510–11.

Cελήνη Τοξότῃ:
Business and Astrology in the Papyri

Giuliano Masola and Nicola Reggiani

Abstract: A private letter on papyrus found in the ancient Egyptian city of Oxyrhynchus (*P.Oxy.* LXV 4483, 194 CE) offers a unique instance of practical use of astrology in ancient times. A deep examination of the text and of the astrological event cited in it (a position of the Moon in Sagittarius, particularly favourable for making deals) will be provided, with references to other documents, ancient horoscopes, and literary texts, in order to show how the study of ancient astrology can prove useful to understand texts and artifacts, and how star-gazing was influential in every aspect of everyday life, as often happens even today.

> *Une autre carte, s'il vous plaît; merci.*
> *Arcane sixième: le Sagittaire.*
> *Vénus transformée en ange ailé*
> *envoie des flèches vers le soleil.*[1]

1. Introduction

Evidence provided by Greek papyri range over every aspect of everyday life in ancient Egypt, and faith in the supernatural is undoubtedly one of

---

This paper is based on the research presented by Dr. Masola at the end of the Papyrology class at the University of Parma (2011/12, Prof. Isabella Andorlini). He is the author of Introduction and Part 2, while Dr. Reggiani wrote Part 3 and the Conclusion. Editions of papyri are cited according to the standard abbreviations given by the online *Checklist of Editions of Greek, Latin, Demotic, and Coptic Papyri, Ostraca and Tablets*, edited by Joshua D. Sosin, Roger S. Bagnall, James Cowey, *et al.*, at [accessed 10 November 2012]:
http://library.duke.edu/rubenstein/scriptorium/papyrus/texts/clist.html

1 Michel Tournier, *Vendredi ou les Limbes du Pacifique* (1969; repr. Paris: Gallimard, 2004), p. 9.

Giuliano Masola and Nicola Reggiani, 'Cελήνη Τοξότῃ: Business and Astrology in the Papyri', *Culture And Cosmos*, Vol. 17, no. 1, Spring/Summer 2013, pp. 93–110.
www.CultureAndCosmos.org

the most interesting issues, in pointing out how even at that time—mostly as nowadays—everyday life was always characterised by an atmosphere of irrationality that perhaps is a basic part of men and comes along with their choices from time immemorial.[2] Franz Cumont, in his thorough research about 'the Egypt of the astrologers', has shown how astrology, in particular, used to permeate very wide parts of common life.[3] Besides a considerable amount of horoscopes and other similar texts, well known and well studied (from customised natal charts to more extensive treatises of practical astrology[4]), which have been defined as 'the physical evidence of

---

2 On methodological and scientific boundaries of Papyrology see, for example, Eric G. Turner, *Papiri greci*, updated Italian edition by Manfredo Manfredi (Roma: Carocci, 2002) [or. ed. *Greek Papyri. An Introduction* (Oxford: Oxford University Press, 1968 and 1980)], and also Roger S. Bagnall, *Reading Papyri, Writing Ancient History* (London and New York: Routledge, 1995).

3 Franz Cumont, *L'Egitto degli astrologi* (Milano: Mimesis, 2003) [or. ed. *L'Égypte des astrologues* (Bruxelles: Fondation Égyptologique Reine Élisabeth, 1937)]. Cumont's work reports any reference to Egyptian society which can be taken from astrological texts, but it can be read backwards as well, that is noting how much everyday aspects had astrological fallouts: compare with Donata Baccani, *Oroscopi greci. Documentazione papirologica* (Messina: Sicania, 1992), p. 50. For a general introduction on ancient astrology see Wolfgang Hübner, 'L'astrologie dans l'Antiquité', *Pallas*, Vol. 30 (1983): pp. 1–24, and Roger Beck, *A Brief History of Ancient Astrology* (Malden-Oxford-Carlton: Blackwell Publishing, 2007); on Egypt, in particular, see Dorian G. Greenbaum and Micah T. Ross, 'The Role of Egypt in the Development of the Horoscope', in Ladislav Bareš et al., eds., *Egypt in Transition. Social and Religious Development of Egypt in the First Millennium BCE. Proceedings of an International Conference (Prague, September 1–4, 2009)* (Prague: Czech Institute of Egyptology, 2010), pp. 146–82. See also, very recently, Glen M. Cooper, 'Astrology: The Science of Signs in the Heaven', in Paul T. Keyser and John Scarborough, eds., *Handbook of Science and Medicine in the Classical World* (Oxford: Oxford University Press, forthcoming), with further bibliography.

4 The reference works on this topic are: Otto Neugebauer and Henry B. van Hoesen, *Greek Horoscopes*, 2nd ed. (1959; repr. Philadelphia: America Philosophical Society, 1987) [in the text it is referred to as *GH*]; Baccani, *Oroscopi*; Alexander Jones, *Astronomical Papyri from Oxyrhynchus (P.Oxy. 4133–4300a)*, 2 vols., (Philadelphia: American Philosophical Society, 1999). See also James Evans, 'The Astrologer's Apparatus: A Picture of Professional Practice in Greco-Roman Egypt', *Journal for the History of Astronomy*, Vol. 35 (2004): pp. 1–44.

astrologers' activity',[5] papyri have left to us a peculiar instance, showing how star observation might be applied to every meaningful moment of personal life and activities.[6]

2. Elis' letter to Karpos: business and astrology

Papyrus 4483 from Oxyrhynchus (*editio princeps* by Alexander Jones and Paul Schubert[7]), one of the many documents found in the excavations of this famous Egyptian city and kept at the Sackler Library at Oxford (inv. P.Oxy. 4483, Trismegistos nr. 78583)[8], bears the nearly complete text of a private letter written by one Elis to a friend of his, Karpos, reminding him about some household effects (namely plates) which have been ordered.[9] It is a very ordinary text, comparable to many other papyri, both published

---

[5] Alexander Jones, 'Astrologers and Their Astronomy', in Alan K. Bowman et al., *Oxyrhynchus: A City and Its Texts* (London: Egypt Exploration Society, 2007), pp. 307–14, here p. 308.

[6] Paul Schubert, *Vivre en Égypte gréco-romaine. Une sélection de papyrus* (Vevey: Éditions de l'Aire, 2000), p. 89.

[7] Alexander Jones and Paul Schubert, 'Letter of Elis to Carpus', in Michael W. Haslam et al., eds., *The Oxyrhynchus Papyri*, LXV (London: Egypt Exploration Society, 1998), pp. 174–75 [in the text the *editio princeps* is referred to as *ed.pr.*].

[8] In general on the site of Oxyrhynchus and its papyrological relevance, see Alan K. Bowman et al., *Oxyrhynchus: A City and Its Texts* (London: Egypt Exploration Society, 2007), and Peter Parsons, *City of the Sharp-nosed Fish: Greek Lives in Roman Egypt* (London: Weidenfeld & Nicolson, 2007). In particular, among papyrus finds, it is worth stressing 'the presence of an entirely unprecedented number of astronomical and astrological papyri' (Jones, *Oxyrhynchus*, p. 308); some of them will be cited below.

[9] The fragment is concisely described also in the 'virtual exhibition' *Oxyrhynchus: A City and His Texts* at the website *Oxyrhynchus Online*, http://www.csad.ox.ac.uk/poxy/vexhibition/4483.html [accessed 10 November 2012], where a photograph is also available. It is commonly thought that the plate order was made by Karpos, but see below for a different hypothesis.

and unpublished,[10] but, just following such an introduction, Elis suggests that Karpos plan a future meeting with a friend of the latter's (the context does not tell us whether there is any relations with the plate order or not) on the ground of a specific astronomical alignment: 'when the Moon is in Sagittarius'.[11]

The papyrus sheet (measuring 9.4 x 16.2cm) is preserved almost completely, except some rectangular holes of different sizes not affecting legibility apart from line 6, which is entirely lost (just few traces survive at the end of the line). Line 1 is very damaged too, yet it is possible to supply the missing letters (see below). Text distribution follows a regular pattern, suggesting a skilled hand, though some mistakes occur, clearly due to typical phonetic phenomena depending on the influence of spoken Greek (ll. 2, 9: simplification of the diphthong ει in ι; ll. 7, 11: exchange between ν and μ; l. 9, gemination of c); there is however one instance of a mistake corrected by the scribe (l. 9: τον corrected in τῶν). The twelve lines of writing run along the fibres of the *recto* in an uncial handwriting ('inelegant but regular capitals'[12]) datable to the second century CE, not sloping and quasi-'literary' in separating the letters one from another, a clear clue of a professional writer, this writing has indeed a little less accurate and elegant *ductus* than *P.Oxy.* III 589 *descriptum*, an official letter of the second century CE, an example of 'good-sized uncial' in the editor's opinion.[13] Line columnation is not always regular; there is an upper margin about 1.5cm high and a left-hand one of about 1cm, while much of the sheet was left blank in the bottom. On the right-hand side, the first three lines leave a blank margin about 1cm large, while the following ones reach the sheet edge, except the last one, containing the closing

---

10 Compare, for example, with *P.Stras.* VI 553, another letter from the second century CE, in which the same beginning structure μ]νήcθητητ[ι] περὶ... ('remember/don't forget…') occurs.

11 On planetary alignments in ancient astrology see Beck, *Astrology*, pp. 23 ff. (part. 23–24 about the Moon).

12 Jones and Schubert, 'Letter', p. 174.

13 Bernard P. Grenfell and Alan S. Hunt, eds., *The Oxyrhynchus Papyri*, III (London: Egypt Exploration Society, 1903), p. 281. Compare with Herman Harrauer, *Handbuch der griechischen Paläographie* (Stuttgart: Hiersemann, 2010), p. 344, nr. 156, table 141.

greetings, separated from the text, slightly indented, and followed by a horizontal filling mark. Figures—represented as usual with alphabetic characters—are regularly supralined. On the *verso*, 'faint smudges may indicate the former presence of the address, along the fibres'.[14]

The document is briefly commented in its *editio princeps* but, besides the two later references cited above, as brief as the first edition, and the mention by Alexander Jones,[15] it does not seem to have raised the interest it should deserve because of its peculiarity. Therefore it may be useful to present a full transcription of the text with a wider and more articulate commentary, as an introduction and a context for a more specific in-depth examination of its astrological features.

P.Oxy. LXV 4483
(Letter from Elis to Karpos)

Oxyrhynchus   *ante* 9 September 194 CE

(see Fig. 1)

Ἧλις Κάρπῳ τῷ τιμιω-
τάτῳ πλ<ε>ῖςτα χαίρειν.
μνήςθητι περὶ τοῦ ἐντο-
λικοῦ τ[ο]ῶν ϲκουτλίων τῶν
5 τριῶν, μεγάλων β̄ καὶ
[.].[c. 11-13]....
ϲύνβαλε τῷ φίλῳ ϲου ϲελή-
νης οὔϲης Τοξότῃ, ὥρ(ᾳ) δ̄· γίνε-
ται ἐκ<ε>ῖ Θὼθ ῑβ, ἐκ<ε>ῖ ἐϲ{ϲ}τιν πά-
10 λιν καὶ τῇ ῑγ καὶ ῑδ / μέχρι ὥρ(αϲ) ζ̄·
ἐν ταύταις ϲύνβαλε τῷ φί-
λῳ ϲου. διευτύχει —

7, 11 *l.* ϲύμβαλε

Elis to his most esteemed Karpos, many greetings. Remember the order of the three plates, two big ones and [one small

---

14 Jones and Schubert, 'Letter', p. 174.

15 Jones, 'Astrologers', p. 313: 'There is a pretty example of this [*sc.* of katarchic astrology: see below] application in a personal letter from the end of the second century, LXV 4483: the writer advises his correspondent to fulfil an engagement with a friend while the moon is in Sagittarius, and he states the precise dates and times bounding this interval, obviously read off an ephemeris [see below]'.

one?]. Meet (or: make a contract with) your friend when the Moon will be in Sagittarius, at the fourth hour: it arrives there on 12 Thoth, and will be still there on 13 and 14, till the seventh hour. Meet (or: make the contract with) your friend during this time. Farewell.

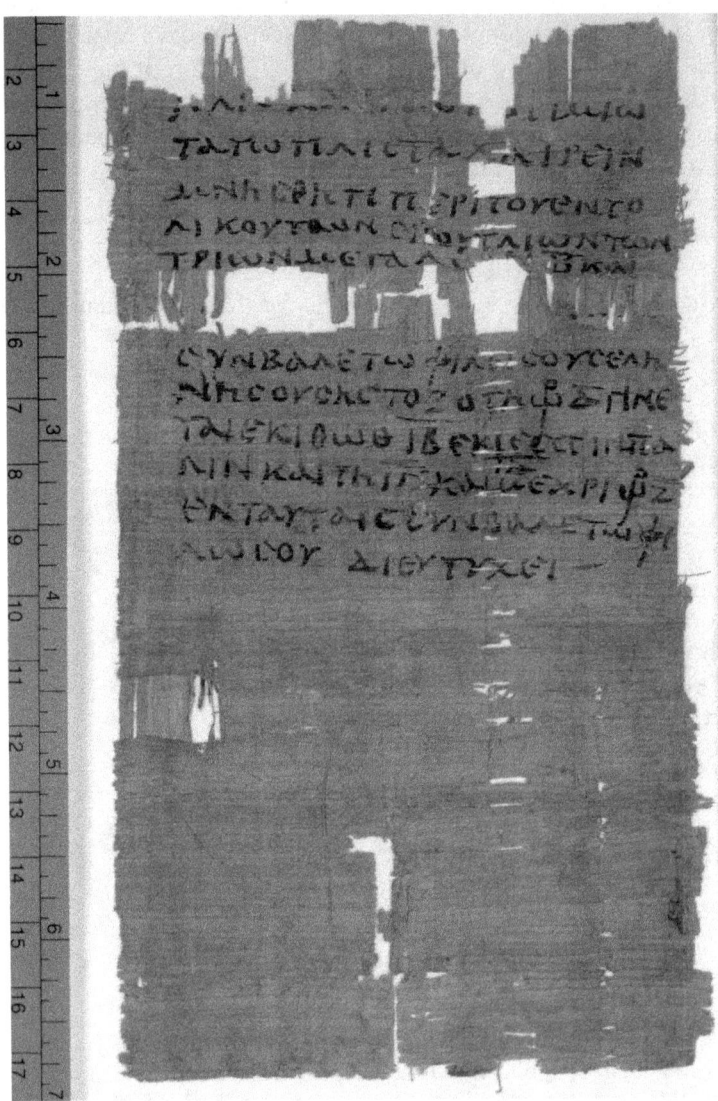

Figure 1 (opposite page): *P.Oxy.* LXV 4483, from *Oxyrhynchus Online*, http://163.1.169.40/gsdl/collect/POxy/index/assoc/HASH7ac1/16aaea29.dir/POxy.

v0065.n4483.a.01.hires.jpg [accessed 7 April 2013]. Image courtesy of the Egypt Exploration Society and Imaging Papyri Project, Oxford.

Commentary

1-2 The friendly style of the opening greetings (τῷ τιμιωτάτῳ πλεῖςτα χαίρειν is a recurrent formula in letters, especially in the first centuries of the Roman age), as well as of the entire letter, makes us suppose that the sender and the addressee used to know each other quite well: it was so natural for Elis to give Karpos a word of advice, here presented in the uncommon form of a 'horoscope'. Moreover, no internal evidence suggest that it was the answer to a previous explicit request by Karpos, but rather a suggestion spontaneously given by Elis, who should have been well versed in astrology (see below).

4 ϲκουτλίων: loan word from Latin *scutula* 'serving dish'.[16] Such a term is seldom attested in documentary papyri of the Roman age, all of them (with the sole exception of our piece) being lists or inventories of tools: *P.Lond.* II 191.10 (103–117 CE: two wooden *skoutlia* for home use); *P.Oxy.* IV 741.29 (second century CE: two *skoutlia*); *P.Mich.* IX 576.6 (third century CE: a big *skoutlion*); *P.Oxy.* XIV 1657.3 (late third century CE: a bronze *skoutlion*, perhaps for military use). They do not seem to have any ritual or magical use at all, though it is possible that they were ordered by Elis the astrologer and not by Karpos as is usually thought (see below).

The o corrected by the scribe is not really crossed out nor erased, but rather overwritten with the ω of τῶν.

6 The *editores principes* already thought that, from the syntax of the sentence, in this line there must have been reference to a third plate (cf. *ed.pr.*, note *ad loc.*), likely described as 'small', with the other two being called 'big'.[17] The traces at the end of the line are interpreted as a horizontal filler mark, similar to the one in l. 12.

It seems very odd that Elis, in reminding Karpos of the three plates,

---

16 See Sergio Daris, *Il lessico latino nel greco d'Egitto*, 2nd ed. (1971; repr. Barcellona: Pontificio Istituto Biblico, 1991), p. 104; *ed.pr.*, n. *ad loc.*

17 See also Schubert, *Vivre*, p. 89: 'et (une petite?)'.

	specified not only their exact number, but also their sizes. If the order was really made by Karpos, it would have been sufficient to recall just the plates. A more articulate reminding may suggest that it was Elis himself who had ordered them, likely through Karpos. In this case, of course, the spontaneous astrological piece of advice would have been less unselfish than presented above.
7	Here the verb *symballō* may bear either the broad meaning of 'meet' or the more specific one of 'make a contract'. As the *editores principes* noted (*ad loc.*), 'this verb can have various meanings which would suit our text. Maybe the addressee should simply meet his friend. On the other hand, we find, in the context of oracular questions, the word σ]υνβαλῖν used as "making a deal" (P. G. M. XXX d3)'. Their indecision, reflected in their translation 'meet (or make a contract with)', has been cleared up by Paul Schubert in favour of the broader sense ('donne rendez-vous' and 'rencontre' in the translation by J.-M. Jacot-Descombes[18]), but the astrological value of the condition the text refers to makes us rather lean towards the more specific, contractual sense (see below), though it is impossible to state whether it was related to the plate order or not.
7–8	*Selēnē Toxotēi* is the standard expression used to refer to the Moon in Sagittarius in astrological texts: for further occurrences and the astrological meaning see below.
8	ὥρᾳ is abbreviated (as ὥρας at l. 10) with the monogram ⟨φ⟩ (ω and ρ overlapping), which characteristically occurs in other astrological texts: compare with, for example, *P.Oxy.Astr.* 4242.2 (*post* 212 CE); 4247.1 (*post* 233 CE); 4271.3 (*post* 354 CE).[19] The same symbol was also used to abbreviate the word ὡροσκόπος ('ascendant'[20]), as in *P.Oxy.Astr.* 4242.3 (*post* 212 CE); 4247.6 (*post* 233 CE); 4263.1 (*post* 299 CE).[21] Such a word is in fact a compound of ὥρα, with the proper meaning of 'observation of the hour', i.e. of

---

18 Schubert, *Vivre*, p. 89.

19 See Jones, *Oxyrhynchus*, p. 61.

20 See Neugebauer and van Hoesen, *Horoscopes*, p. 7.

21 See Jones, *Oxyrhynchus*, p. 62.

the star positions at the moment of a person's birth, or of any other important event.[22]

8–11   Elis advises Karpos to meet his friend in some time between the fourth hour of day 12 and the seventh hour of day 14 of the month of Thoth (29 August–27 September, according to the traditional Egyptian calendar): this corresponds to 9–10–11 September 194 CE (compare with *ed.pr*, n. *ad loc.*, as confirmed also by the astronomy journal *Coelum*[23]; see Fig. 2), and lets us date the papyrus exactly.[24] Syntax is doubtful here, so that the sentence might also seem to refer to an antelucan meeting (cύνβαλε τῷ φίλῳ cου Cελή|νηc οὔcηc Τοξότῃ, ὥρ(ᾳ) δ̄), but it is evident that propitious is the entire lapse of time during which the Moon is in Sagittarius, till the seventh hour of 14 Thoth. Of course the time computation of heavenly positions has been localized to Oxyrhynchus.[25]

The day number 14 (ιδ) is inserted above the line due to a banal omission by the scribe.

12   The closing farewell formula (διευτύχει) is typical of letters and petitions, but widely occurs also in personal horoscopes: see *P.Oxy.Astr.* 4249; 4264; 4266 (see *ed.pr.*, n. *ad loc.*).

---

22 See Evans, 'Apparatus', p. 1, on the 'horoscopic' nature of Greek astrology; Cooper, 'Astrology', pp. 9–10, on Ptolemaic Egypt as the 'birthplace of Horoscopic Astrology'. On hour calculation in astronomy/astrology see Jones, *Oxyrhynchus*, pp. 14–15.

23 Aldo Vitagliano, 'Una risposta e una lectio magistralis sulla posizione della Luna', *Coelum*, Vol. 16, no. 160, (June 2012): pp. 77–78 (answer to the question sent by Giuliano Masola).

24 On the calculation of dates see Neugebauer and van Hoesen, *Horoscopes*, pp. 1–2; Jones, *Oxyrhynchus*, pp. 49 ff.

25 On the calculation of the positions of celestial bodies see Baccani, *Oroscopi*, pp. 66–77; Jones, *Oxyrhynchus*, pp. 15 ff. (on the Moon, in particular, pp.18–19).

102 Σελήνη Τοξότη: Business and Astrology in the Papyri

Figure 2: The Moon in Sagittarius in the morning of September 10, 194 CE, according to a modern astronomy software ('Stellarium 0.11.3'; for any reference see http://www.stellarium.org [accessed 7 April 2013]).

3. The Moon in Sagittarius: An Astrological Survey

The astronomical event cited in the letter, the apparent passage of the Moon through the zodiacal constellation of the Sagittarius (Archer),[26] clearly perceived as particularly favourable, is an interesting starting point for an original *excursus* through the astrological references provided by

---

26 On this constellation in ancient times, see Franz Boll, *Sphaera: Neue griechische Texte und Untersuchungen zur Geschichte der Sternbilder, mit einem Beitrag von K. Dyroff* (Leipzig: Teubner, 1903), pp. 130–31 and 181–208; Richard H. Allen, *Star Names. Their Lore and Meaning*, 2nd ed. (Mineola: Dover, 1963) [1st ed. *Star-Names and Their Meanings* (New York and London: Stechert, 1899)], pp. 351–57; Alfredo Cattabiani, *Planetario. Simboli, miti e misteri di astri, pianeti e costellazioni* (Milano: Mondadori, 1998), pp. 202–9. On its Mesopotamian origins: Giovanni Pettinato, *La scrittura celeste. La nascita dell'astrologia in Mesopotamia* (Milano: Mondadori, 1999), p. 98; Gennadij Kurtik and Alexander Militarev, 'Once More on the Origin of Semitic and Greek Star Names: An Astronomic-Etymological Approach Updated', *Culture and Cosmos*, Vol. 9 (2005): pp. 3–43, here pp. 25–26.

Graeco-Roman papyri from Egypt and other ancient sources. In fact, after a search hampered by the lack of most of the astrological texts in the available electronic databases (due to the literary or paraliterary nature of such texts), we noted that the occurrences of the condition 'Moon in Sagittarius' are not common, being mentioned in just six more papyri:

P.Aberd. 13 (post 187 CE): Cελήνη νε Δοξ[ότη] (l. Τοξ[ότη]) [= GH 187];
P.Oxy.Astr. 4242.7 (post 212 CE): Cελήνη Τοξότη [Jones, Oxyrhynchus, I 256–7 e II 378–9];
P.Oxy. XXXVI 2790v.i.6 (post 255-57 CE): Cελήνη Τοξότη [= Baccani, Oroscopi, nr. 15];
P.Oxy. XII 1565.7 (post 293 CE): Cελήν[η] Τοξότη [= GH 293.VIII];
P.Oxy. XL 3298.i.7 (3rd cent. CE, 2nd half): Cελήνη Τοξότη μοιρ(ῶν) γ [= Baccani, Oroscopi, nr. 14];
P.Oxy. XVI 2060.5 (post 478 CE): [Cε]λήνη Τοξότη μο(ιρῶν) ιζ λε(πτῶν) ι [= GH 478].

It is to be added to them *P.Lond.* I 98 (post 95 CE) [= GH 95], an elaborate horoscope joined to the astrological treatise in Greek and Coptic written on the *recto* of the papyrus on the *verso* of which Hyperides' famous funeral oration was transcribed:[27] here the position of the Moon is lost at lines 22 ff. of the first column, but it has been possible to infer its position in Sagittarius thanks to the positions of the other planets mentioned.[28]

Even in some Demotic horoscopes there are references to the Moon in Sagittarius (*i'ḥ n p3 nty 3tḥ*);[29] however, all of these cases concern personal

---

27 On Hyperides' speech and the papyrus containing it see Judson Herrman, ed., *Hyperides: Funeral Oration* (Oxford: Oxford University Press, 2009).

28 Neugebauer and van Hoesen, *Horoscopes*, pp. 34–35.

29 The reference is to two *ostraka* (Herbert Thompson, 'Demotic Horoscopes', *Proceedings of the Society of Biblical Archaeology*, Vol. 34 (1912): pp. 227–33, nr. 1 [post 18 CE], and *O.Stras.Dem.* 270 [post 35 CE]) described and published by Otto Neugebauer, 'Demotic Horoscopes', *Journal of the American Oriental Society*, Vol. 63 (1943): pp. 115–27, nos. 3 and 4. *i'ḥ* is the Demotic name for the Moon, and *p3 nty 3tḥ* is the Sagittarius, 'he who stretches (the bow), he who strikes' (see Wilhelm Spiegelberg, 'Die ägyptischen Namen und Zeichen der

horoscopes, that is calculations of what is nowadays called a 'natal chart': the (often graphic) description of the heavenly state at the moment of a person's birth, as observed in the place in which it happened, to make predictions and gather information about that person's life,[30] just as in the curious inscription scratched by one Artemidoros who went on pilgrimage to the funerary temple of Pharaoh Sethis I at Abydos, active until 359 CE as an oracular sanctuary, and traced (between the second half of the fourth century and the beginnings of the fifth century CE) his own 'natal chart', together with a health wish and an invective against potential erasures. Among the various combinations present in this chart, there is again Τοξότης· | Σελήνη.[31]

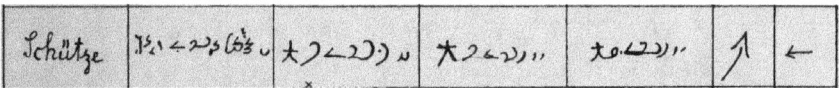

Figure 3: Names and symbols of the Sagittarius in Demotic texts [from Spiegelberg, 'Tierkreisbilder', Table IV].

Tierkreisbilder in demotischer Schrift', *Zeitschrift für Ägyptische Sprache*, Vol. 48 (1911): pp. 146–51, here p. 148, no. 9; see Fig. 3); later, in Coptic, it will become the 'arrow' *tout court*, ⲥⲟⲧⲉ < *sty.y*: compare with Sergio Pernigotti, 'Una rilettura del P.Mil.Vogl. Copto 16', *Aegyptus*, Vol. 73 (1993): pp. 93–125, here pp. 112–13.

30 See Baccani, *Oroscopi*, pp. 39–48; Jones, *Oxyrhynchus*, p. 4; Jones, 'Astrologers', pp. 308–9. Such attestations are similar to the literary ones: in particular, the Moon in Sagittarius occurs four times in the work of Vettius Valens, a Roman astrologer who wrote *Anthologiae* around 152–162 CE. The attestations are: II 21, p. 86, 10–16; VII 2, p. 268, 7–15 and p. 269, 11–28; VII 5, p. 281, 24–p. 282, 2 (ed. Pingree) [= *GH* L105; L114.IX; L117.VI; L134.VI]. On Vettius Valens see Boll, *Sphaera*, pp. 59–72; Cooper, *Astrology*, p. 5. It is noteworthy also the representation of the Moon in Sagittarius painted into the 'Zodiac Tomb' at Athribis, in the middle of the second century CE: see William M. Flinders Petrie, *Athribis* (London: School of Archaeology in Egypt, 1908), p. 12, and see Fig. 4.

31 *I.Memnoneion* 641 = Paul Perdrizet and Gustave Lefebvre, *Inscriptiones Graecae Aegyptii*, III (Nancy: Berger-Levrault, 1919), no. 641 = *GH* 353; see Ian Rutherford, 'The Reader's Voice in a Horoscope from Abydos', *Zeitschrift für Papyrologie und Epigraphik*, Vol. 130 (2000): pp. 149–50, and, on the context, Ian Rutherford, 'Pilgrimage in Greco-Roman Egypt: New Perspectives on Graffiti from the Memnonion at Abydos', in Roger Matthews and Cornelia Roemer, eds., *Ancient Perspectives on Egypt* (London: Routledge, 2003), pp. 171–89.

Figure 4: The Moon in Sagittarius in the 'Zodiac Tomb' at Athribis [middle second century CE; adapted from http://www.digitalegypt.ucl.ac.uk/ athribis/tomb.html [accessed 7 April 2013; compare to Petrie, *Athribis*, Pl. XXXVII]. Copyright Petrie Museum of Egyptian Archaeology UCL.

Nevertheless, all of these instances are very different from the one provided by Elis' letter, in which the astrological fact was used to pick out the most convenient moment to perform a certain activity or to make a certain decision.[32] Therefore the text is not retrospective,[33] but projected on a forthcoming event (though likely in the same year the letter was

---

32 Different is the case of *CCAG* 1, p. 103.1–30 (*post* 474 CE) [= *GH* L475], in which the Moon in Sagittarius is used just to determine the exact time of a ship's arrival ('I said when the Moon would be in Sagittarius or Pisces the ship would arrive. And when the Moon came into Sagittarius they arrived'). The abbreviation *CCAG* stands for *Catalogus Codicum Astrologorum Graecorum*, 12 vols. (Bruxelles: Lamertin, 1898–1953), a collection of medieval and Renaissance manuscripts containing Greek astrological texts.

33 See Jones, *Oxyrhynchus*, p. 57.

composed, since the writer did not feel the need to state a reference year): one of the so-called 'katarchic' horoscopes.[34]

Elis must have been an astrologer,[35] since he uses the monogram ΩP to abbreviate the word *hōra*—as we have seen above, this was a very common symbol in personal horoscopes produced by such professional exegetes of stars. Further on, as an astrologer he must have been in possession of an almanac, or astronomical calendar (*ephemeris*) registering, among other data, the position of the moon (in terms of sign of the zodiac and degrees) day by day (usually at sunset), in addition to the exact moment of transition from a sign to the following one.[36] So he was not unaware of the influences that, according to ancient theories, the heavenly bodies might have on human activities, and that many astrological treatises used to report for the use of experts.[37] It is just one of such 'handbooks', a part of which (concerning the Moon) has been fortuitously included in magical papyrus *PGM* I 3 (ll. 275–81; the papyrus, roughly dated to the fourth century CE, is inventoried as no. 2391 in the Museum of Louvre, and is also called 'Mimaut Papyrus' from the name of its former owner) because of the magical implications suggested for each

---

34 See Beck, *Astrology*, pp. 10 and 101; Jones, 'Astrologers', pp. 308 and 313; Cooper, *Astrology*, p. 3. The name comes from *katarchai*, i.e. 'inaugurations' of activities: such horoscopes provided the basis for individual decision-making. A literary example is provided by *GH* L484, recalling a horoscope calculated with reference to Leontius' usurpation of the imperial throne (see Beck, *Astrology*, p. 95), but see also Petronius, *Satyricon*, 30.4.

35 This is the opinion expressed also in the page about this papyrus in the web site *Oxyrhynchus Online* cited above. For an interesting *excursus* on the practice of an astrologer's work in Graeco-Roman Egypt see Evans, 'Apparatus'.

36 See Jones, *Horoscopes*, pp. 40–42; Jones, 'Astrologers', pp. 310–13. Such tables are attested in papyri: as regards chronology, the closest to our letter is *P.Oxy.Astr.* 4181, dated to 161 CE. It is interesting the table of Babylonian tradition published in Otto Neugebauer, 'A Babylonian Lunar Ephemeris from Roman Egypt', in E. Leichty et al., eds., *A Scientific Humanist: Studies in Memory of Abraham Sachs* (Philadelphia: Kramer Fund, 1988), pp. 301–4.

37 On the astrological treatises attested in the papyri see Otto Neugebauer and Henry B. van Hoesen, 'Astrological Papyri and Ostraca: Bibliographical Notes', *Proceedings of the American Philosophical Society*, Vol. 108 (1964): pp. 57–72, and Baccani, *Oroscopi*, pp. 29–37.

astrological position, that reveals to us what was, for an ancient astrologer, one of the meanings of the Moon in Sagittarius:[38]

[κύκλος]· **Cελήνη** ἐν [. . . . . . . ἢ Π]αρθέν|ῳ
πανάλ[ωτον, π]οίει λεκαν[ομαντεία]ν, ὡς θέ[λε]ιc, ἐν
[Καρκίνῳ πα]ραιτίαν, | ἀερομαντεῖο[ν . . . .]c, ἐν
Διδύμο[ιc καθά]μματα [. . . . . . ἐν] Ζυγῷ |
πρόcκλ[ηcιν . . . .]νων ἀπόλυ[cιν . . . . .
ν]εκυομαντ[είαν], | ἐν Ἰχθ[ύcιν . . . .]οιω ἢ
ἀγώγ[ιμον], **ἐν Τοξ[ότῃ ἐπι]τηδίαν**, || . . . .
. . ἐν Αἰ]γοκέρῳ ἰκα[νόν . . . . . .] | ἐν [

Horoscope: **Moon** *in ... or in Virgo: anything is obtainable, perform bowl divination, as you wish; in Cancer: perform the spell of reconciliation, air divination...; in Gemini: perform spells of binding ... ; in Libra: perform invocation ... spell of release ... necromancy; in Pisces: ... or love charm;* **in Sagittarius: (conduct) business** *...; in Capricorn: do what is appropriate ...; in ...*

The moment in which Moon was in Sagittarius was therefore considered as the most propitious for making deals (ἐπιτηδ<ε>ίαν: 'Geschäft' in the German translation, 'business' in the English one; literally 'the necessary works'[39]). It is curious noting (*nihil sub sole novi...*) that for contemporary astrology such a condition still has a beneficial influence—among other things—on business planning and investments.[40]

---

38 Karl Preisendanz, ed., *Papyri Graecae Magicae. Die griechischen Zauberpapyri*, 2 vols., (Leipzig and Berlin: Teubner, 1928), 1: pp. 30–63; the passage is quoted from p. 44 [German translation at p. 45, English version in Hans D. Betz, ed., *The Greek Magical Papyri in Translation, Including the Demotic Spells* (Chicago: University of Chicago Press, 1986), p. 26, here followed with minor changes].

39 The word usually occurs in the papyri as a neuter noun; but compare this with *P.Oxy.* XVI 1899, 15 (476 CE): ἐπιτηδ<ε>ίαν 'serviceable'.

40 This piece of information is purely *exempli gratia*, and has been taken from some today's internet sites providing horoscopes and astrological services. The Sagittarius' link to business deals (in particular 'esteri' and 'legali', foreign and legal deals) is further confirmed, for example, by Ursula Lewis, *Il tuo oroscopo*

Maybe it derives from the character attributed to the *Toxotēs*: a sign devoted to adventures and risks, purely masculine—as another interesting astrological papyrus reports, 'a masculine sign with a double body'[41] (ἐν ζ]ῳδίῳ ἀππεν[ι]κῷ δicώμῳ Τ[ο]ξό[τη: *P.Oxy.Astr.* 4245.5, dated to 218 CE).

Of course that was not its only value: a Demotic astrological text of Roman age about the influences of the star Sothis/Sirius (*n3 sḥnyw Spdt*) explains that if it rises when the Moon is in Sagittarius (*i-ir[-s] ḫ ' r i'ḥ ḫn p3 <nty> 3tḥ*) it will favour (as far as we can understand from the lacunose text) grain in the fields, something in the country of the Syrians, death, abundance of 'weakness' by night and day, and something else being 'filled' or—in line with what we have seen above—'paid' (*prt n t3 sḥt [. . . . . . .*] | *ḥr p3 tš n p3 Iḥwr r mwt . . . r ḫpr r . . .ny r 'š3 n [gby] n grḥ mtry [r . . . .] r | šm . . . mtw-f mḥ r-bnr*).[42] In the magical papyrus *PDM Suppl.*168–84, the days when the Moon is in Sagittarius (but also in Leo, Aquarius or Virgo) are instead favourable for reciting a formula of 'arrival of the god',[43] just as in *PGM* I 5.380 the Moon in Sagittarius (or Aries, Leo, Virgo) helps to invoke Hermes' apparition; in another brief magical

---

(La Spezia: Fratelli Melita Editori, 1994), p. 12, a common 'handbook' of modern 'practical astrology'.

41 On the double nature of Sagittarius see Ptolemy, *Tetrabiblos,* I.11, p. 67 Robbins; Boll, *Sphaera*, pp. 181–82 (διπρόcωπον, διφυέc); Beck, *Astrology*, p. 55. In ancient times such a doubleness was represented not only with the composition of a human and a horse body, but also with the double head, human on one side and lion-like on the other one.

42 P. Cairo inv. 31222.1–3: see George R. Hughes, 'A Demotic Astrological Text', *Journal of Near Eastern Studies*, Vol. 10 (1951): pp. 256–64, part. pp. 258–60. For *mḥ r-bnr* with the meaning of 'be full, filled, paid', compare with ibid., p. 260, n. 10. Other various interpretations of this position, not relevant for us, can be found in Babylonian horoscopes: see for example, Pettinato, *Scrittura celeste*, pp. 119–20 and 357.

43 Betz, *Magical Papyri*, p. 329 (translated by Janet H. Johnson); text edition in Janet H. Johnson, 'Louvre E 3229: A Demotic Magical Text', *Enchoria*, Vol. 7 (1977): pp. 55–102, here pp. 73–74 (vi.25–7; vii.1–14). Such formulas were likely used to propitiate visions of gods (see also ibid., pp. 90–91), in this case Imhotep.

lunar 'horoscope' (*PGM* II 7.284–99) it is just said that the Moon in Sagittarius favours invocations or spells to the Sun and the Moon.[44]

From the occult side of this matter we are brought in the world of agricultural works by the more practical Latin authors: they inform us that the nights of (full) Moon in Sagittarius (or Leo, Scorpio, Taurus) are the most suitable to prune vines in order to avoid rats and mice eating them: so wrote Columella (*De arboribus* 15.1[45]), followed by Pliny (*Naturalis historia* XVII.215), who however does not mention mice, referring instead to a general good growth of the plant. Upon this 'remède magique préventif, non retenu dans le corpus ni ailleurs que chez Pline',[46] Raymond Billiard, recalling Varro, ironically comments: 'Columelle, le sérieux Columelle, ne trouvait rien de mieux à faire en pareille circonstance, que de tailler la vigne à la clarté de la lune, quand cet astre était dans le signe du Lion, du Scorpion, du Sagittaire ou du Taureau. Mais peu confiants dans cet expédient astronomique, les insulaires de Pandataire, qui décidément étaient bien mal partagés, s'ingéniaient à dresser dans leur champs mille souricières et pièges de tous modèles'.[47]

---

44 According to Claudius Ptolemy (*Tetrabiblos*, IV.4.182, p. 390 Robbins) the Moon in Sagittarius (and Pisces) would favour necromancy and evocation of demons.

45 *Vites, quae secundum aedificia sunt, a soricibus aut muribus infestantur. Id ne fiat, plenam lunam obseruabimus, cum erit in signo Leonis uel Scorpionis uel Sagittarii uel Tauri et noctu ad lunam putabimus*; see Raoul Goujard, ed., *Columelle: <Les arbres>* (Paris: Les Belles Lettres, 1986), p. 9: 'les trois courts chapitres 13, 14 et 15, donnant des recettes parfois magiques (chap. 15) pour protéger la vigne, ne sont pas repris dans le corpus [i.e. the arrangement of Columella's *De re rustica* during the Renaissance], ce qui témoigne d'une conception plus rationnelle de la viticulture'.

46 Goujard, *Columelle*, p. 114. Pliny cites the Moon in Sagittarius two more times in his *Naturalis historia*: in II.78, speaking of cosmology, he notes that our satellite never comes in conjunction with the Sun in that constellation, and in XXX.97, while widely criticizing traditional and magical therapies, he mentions a 'prescription' ordering to oil people suffering of fever with bat wings crushed in oil when the Moon crosses Sagittarius.

47 Raymond Billiard, *La vigne dans l'antiquité* (Lyon: Laffitte, 1913), p. 396; the reference is to *Res rusticae* I.8, 7–9.

4. Concluding remarks

Beside the many possible magical/apotropaic effects of the astrological phenomenon we are dealing with, the 'economic' influence of the Moon in Sagittarius supports, among the different meanings of the verb cύνβαλε used in the letter, the specific sense of 'make a contract' (see above). This is thus an interesting case in which (as usually happens) the cultural context of a document may help to clarify some interpretation problems of the text itself. Moreover, although it is not possible to ascertain whether Karpos' meeting was linked to the plate order or not, his friend's word of astrological advice aimed at letting him (or himself!) strike a bargain. We will never know the end of the story, but this document represents 'a rare, if not unique, instance of a nontechnical document in which data of a kind found in many astronomical texts and tables on papyrus are applied to a practical situation in real life',[48] as a further confirmation of the importance of knowing the history of every aspect of ancient civilizations, to better understand a past that we can still feel, *mutatis mutandis*, as actual.

---

48 Jones and Schubert, 'Letter', p. 174.

Research Note: *Weltall, Erde, Mensch* and Marxist Cosmology in East Germany

Reinhard Mussik

Abstract: This research note discusses the most published book in the German Democratic Republic (GDR): *Weltall, Erde, Mensch* (Space, Earth, Human Being), in terms of the Marxist cosmology, which permeates all chapters of the book and is interwoven on two different levels.

In May 1974, I became the proud owner of a copy of the book *Weltall, Erde, Mensch* (Space, Earth, Human Being) which Walter Ulbricht (1893–1973)—the East German head of state from 1960–1971—called: 'the book of the truth'.[1] Not only '*a* book of truth' but '*the* book of truth'! Like most East-German teenagers, I never really read it.[2] But I liked the gorgeous fold-out pictures, which depicted our solar system, prehistoric plants and animals, stone-age cave art, Egyptian reliefs, medieval labours of the months, the misery of steelworkers in the capitalist society, the October Revolution in Russia, the modern world of socialism and the communist future—replete with giant atomic power plants and scientific stations on the moon.[3]

---

1 'Dieses Buch ist das Buch der Wahrheit', *Weltall, Erde, Mensch: Ein Sammelwerk zur Entwicklungsgeschichte von Natur und Gesellschaft*, 19th ed., (Berlin: Verlag Neues Leben, 1971), p. 5. All translations my own.

2 'Ob es auch das am meisten gelesene Buch war, sei dahingestellt'. In: Stefan Wolle, *Aufbruch in die Stagnation: Die DDR in den Sechsiger Jahren* (Bonn: Bundeszentrale für Politische Bildung, 2005), p. 99.

3 *Weltall, Erde, Mensch*, 22nd ed., (1974), fold-out drawings without page numbers; Wolle, *Aufbruch in die Stagnation*, p. 99.

Reinhard Mussik, 'Research Note: *Weltall, Erde, Mensch* and Marxist Cosmology in East Germany', *Culture And Cosmos*, Vol. 17, no. 1, Spring/Summer 2013, pp. 111–17.
www.CultureAndCosmos.org

Research Note: 'Space, Earth, Human Being'

Figure 1: Cover of *Weltall Erde Mensch* (1974)

I was given this book at my 'Jugendweihe' ceremony (youth consecration) on the stage of the clubhouse in my hometown, Ludwigsfelde. The 'Jugendweihe' was a secular coming of age ceremony for the fourteen-year-olds in the GDR. This tradition was the atheists' answer to religious confirmation and is based in the secular movements of the middle of the nineteenth century. In the 1950s it became widespread in the GDR, where the government encouraged atheism. Between 1954 and 1974 all participants of the 'Jugendweihe' recieved the book *Weltall, Erde, Mensch* as a present. Just to show how ubiquitous this text became: In 1972 more than 95% of the teenage population participated in the 'Jugendweihe'.[4] With approximately four million copies, it was the most

---

[4] Torsten Morche, *Weltall ohne Gott, Erde ohne Kirche, Mensch ohne Glaube: Zur Darstellung von Religion, Kirche und 'wissenschaftlicher Weltanschauung' in Weltall, Erde, Mensch zwischen 1954 und 1974 in Relation zum Staat-Kirche-Verhältnis und der Entwicklung der Jugendweihe in der DDR* (Leipzig: Kirchhof & Franke, 2006), p. 57.

printed book in the GDR.⁵ My copy was from the twenty-second, and final, edition.

The first editions of *Weltall, Erde, Mensch* were inspired by a Czechoslovakian exhibition titled 'The development of Space, Earth and Human Being', and they are adorned with coloured folding-outs, featuring reproductions of paintings by the Czech painter and book illustrator Zdeněk Burian (1905–1981).⁶ The book contained sections from 'German scientists from the East and the West', like the future dissident Robert Havemann (1910–1982).⁷ Walter Ulbricht—at this time first secretary of the Socialist Unity Party and Deputy Prime Minister of the GDR—wrote the preface. He emphasized the significance of the book in the fight against 'superstition, mysticism, idealism, and other non-scientific views'.⁸

Such 'Superstitions' are clearly named in the book. Astrology, for example, is defined as 'a false doctrine, which claims that the fate of people is predestined by the constellations of heavenly bodies and astrology hence distracts people from shaping their own destiny autonomously'.⁹ A definition of 'idealism' is absent in the glossary at the end of the book. Hints of 'religion', 'church' and all related terms are missing as well.¹⁰ The very absence of these terms in the glossary points to the presence of the Marxist tone of the text itself: people familiar with Marxist terminology already know that a fight against idealism must include a fight against the church and religion.¹¹

---

5 Wolle, *Aufbuch in die Stagnation*, p. 98.
6 *Weltall, Erde, Mensch*, 4th ed., (1956), pp. 5–10.

7 Ibid., pp. 7, 10.

8 'Gleichzeitig wird der Kampf gegen Aberglaube, Mystizismus, Idealismus und alle anderen unwissenschaftlichen Anschauungen geführt', *Weltall, Erde, Mensch*, p. 7.

9 'Astrologie: Sterndeutung; Irrlehre, die die Vorbestimmung des Menschen-schicksals aus der Stellung der Himmelskörper behauptet und damit die Menschen abhält, ihr Schicksal selbsttätig zu gestalten', *Weltall, Erde, Mensch*, p. 541.

10 *Weltall, Erde, Mensch*, pp. 540–70.

11 'Idealismus: Bezeichnung für alle philosophischen System und Anschau-ungen, die das *Bewußtsein* (gleichgültig in welcher Form) für das Primäre, das Grundlegende, das Bestimmende gegenüber der *Materie* erklären. ', Manfred Buhr

Over the years, the authors of the book's sections changed and the fold-out pictures by Zdeněk Burian were replaced by drawings by the East German book illustrator Eberhard Binder-Staßfurt (1924–2001), reproductions of photographs, diagrams and maps. But the structure of the book always stayed the same. When I got my copy of the book in 1974, Walter Ulbricht had been dead for almost one year, and his foreword was not printed anymore.[12]

After 1966, and especially after Ulbricht's death, as the relations between the state and church began to relax in the GDR, *Weltall, Erde, Mensch* was no longer kept up to date.[13] In the eyes of the East German leaders, a special 'book of truth' designed to compete with the Bible no longer seemed necessary.[14] From 1975 to 1982 the youth in the GDR got a new book—*Der Sozialismus, Deine Welt* (The Socialism, Your World). And from 1982 until the dissolution of the GDR in 1989, they received the book *Vom Sinn unseres Lebens* (About the meaning of our life) as a present at their 'Jugendweihe'.[15] But only *Weltall, Erde, Mensch* propagated a closed and all-embracing presentation of Marxist cosmology.

The first part of the book—'Space'—is dedicated to cosmology, which Norriss Hetherington has defined as: 'the science, theory or study of the universe as an orderly system, and of the laws that govern it; in particular a branch of astronomy that deals with the structure and evolution of the universe'.[16] This 'Space' section consists of three chapters about 'The Conquest of the Atom', 'The Structure of the Universe' beginning with a history of Astronomy, and 'Space Exploration Breaking new Ground'. The explanation of the structure of the universe in this collected edition is based

---

and Alfred Kosing, *Kleines Wörterbuch der Marxistisch-Leninistischen Philosophie* (Berlin: Dietz Verlag, 1982), p. 155.

12 *Weltall, Erde, Mensch*.

13 Morche, *Weltall ohne Gott, Erde ohne Kirche, Mensch ohne Glaube*, pp. 54–60).

14 Ibid., p. 57.

15 *Der Sozialismus, Deine Welt* (Berlin: Verlag Neues Leben, 1975); *Vom Sinn unseres Lebens* (Berlin: Verlag Neues Leben, 1983).

16 Norriss S. Hetherington, *Cosmology: Historical, Literary, Philosophical, Religious, and Scientific Perspectives* (New York: Garland Pub., 1993), p. 116.

on the theory of the soviet astronomer Victor Amazaspovich Ambartsumian (1908–1996) who claimed that we live in an eternal universe with expanding galaxies.[17]

Ambartsumian was one of the leading Soviet astronomers. He founded a school of astronomy in Armenia and the Byurakan Observatory. As a member of the Supreme Soviet he was one of the most powerful scientists of his time in the Soviet Union. Regardless of his high position in the Soviet government, he was elected as a Member of the Royal Society in 1969! According to him, most stars were born in unbound expanding associations rather than in bound clusters, and he concluded that stars were born explosively. He was the first astronomer to emphasize the importance of activity and explosions in the nuclei of galaxies.[18]

After 1930, the amateur astronomy journal *Mirovedenie* became the organ for the introduction of dialectical materialism into all aspects of astronomical research in the Soviet Union.[19] Every cosmological theory was being put to the test of Marxism-Leninism now. While astronomical research and the development of cosmological theories were never officially restricted in the Soviet Union, all astronomers and astrophysics learned which kind of theories were accepted by the Communist Party and which were not. At least 29 Soviet astronomers—between them a quarter of the scientific staff of the famous Pulkovo Observatory where Ambartsumian researched at this time—were arrested between 1936 and 1937.[20] Ambartsumian was denounced as a cleverly masked enemy of Marxism-Leninism in 1938. Unlike a lot of his colleagues, he escaped persecution and spent the following years in the less important Simeis station. In 1940 he became a member of the Communist Party and began his political career.[21]

---

17 D. Lynden-Bell and V. Gurzadyan, 'Victor Amazaspovich Ambartsumian. 18 September 1908–12 August 1996', *Biographical Memoirs of Fellows of the Royal Society*, Vol. 44, (1998): p. 22.

18 Ibid.

19 Robert A. McCutchoen, 'The 1936–1937 Purge of Soviet Astronomers', *Slavic Review*, Vol. 50, no. 1, (1991): p. 103.

20 Lynden-Bell and Gurzadyan, 'Victor Amazaspovich Ambartsumian', p. 25; McCutcheon, 'The 1936–1937 Purge', pp. 100, 111.

21 Lynden-Bell and Gurzadyan, 'Victor Amazaspovich Ambartsumian', p. 26.

Like many of the surviving astronomers and astrophysicists, Ambartsumian had to develop a cosmology conform to the theory of Marxism-Leninism. According to Maxim W. Mikulak, the Soviet astronomers considered the traditional concept of the infinite universe as progressive while the concept of the finite expanding universe—related to the theory of relativity—should be a modern version of a reactionary, mystical and idealist picture of the cosmos.[22] Soviet astronomers had to accept that the universe is infinite in time and space. In the light of this ideological framing, they avoided the construction of cosmological models, particularly models of the universe as a whole.[23] The 'Space' section of *Weltall, Erde, Mensch* adheres closely to this Soviet trend.

But the term 'cosmology' can be seen in a broader sense too. Fiona Bowie defines it as 'a theory or conception of the nature of the universe and its workings, and of the place of human beings and other creatures within that order'.[24] Marx's social universe is based, as Micheal Neary has stated, 'on value, which he clearly considered in cosmological terms'.[25]

Weltall, Erde, Mensch was also inspired by the book *Die Welträtsel* (The Riddles of the Universe) by Ernst Haeckel (1834–1919).[26] Haeckel presented not only a compendium of modern sciences, but he also tried to refute the dogmata of the church. He stated, for example, that the immortality of the soul is a dogma which stands in unsolvable opposition

---

22 McCutcheon, 'The 1936–1937 Purge', pp. 167.

23 Maxim W. Mikulak, 'Soviet Cosmology and Communist Ideology', *The Scientific Monthly*, Vol. 81, no. 4, (1955): p. 46; W. A. Ambarzumjan and W. W. Kasjutinski, 'Metagalaxis und Weltall', in *Struktur und Form der Materie: Dialektischer Materialismus und moderne Naturwissenschaft* (Berlin: VEB Deutscher Verlag der Wissenschaften, 1969).

24 Fiona Bowie, *The Anthropology of Religion: An Introduction* (Oxford: Blackwell, 2006), p. 108.

25 Michael Neary, 'Travels in Moishe Postone's Social Universe: A Contribution to a Critique of Political Cosmology', *Historical Materialism*, Vol. 12, no. 3, (2004): p. 240, fn. 2.

26 Ernst Haeckel, *Die Welträtsel* (1899; repr. Hamburg: Nikol Verlagsgesellschaft mbH, 2009).

to modern sciences.[27] However, such direct confrontation with the dogmata of the church cannot be found in *Weltall, Erde, Mensch*. The atheist propaganda in the GDR was more subtle. But the idea behind both texts was the same: to spread atheist propaganda with a book containing a closed cosmology based on modern scientific knowledge. Following the logic of 'dialectical and historical materialism' the *Weltall, Erde, Mensch* develops a systematic explanation of the universe, from the birth of our galaxy up to the inevitable global victory of communism.[28] In this sense, the book presents a consistent cosmology, and, as an emblematic text of the former German Democratic Republic, it deserves further research.

---

27 Ibid., p. 267.
28 *Weltall, Erde, Mensch*, pp. 487–503.

NOTES ON CONTRIBUTORS

Clifford J. Cunningham is with The University of South Queensland, and the National Astronomical Research Institute of Thailand. He has undergraduate degrees in science and classical studies from the University of Waterloo in Canada. His prime interest in the history of astronomy is the early detection and study of the first four asteroids. His first book, *Introduction to Asteroids*, was published in 1988. In addition to authoring a four-volume work on asteroid history, he is editor of *The Collected Correspondence of Baron Franz Xaver von Zach*, of which seven volumes have been published by 2013. He has been the history of astronomy columnist for *Mercury* magazine since 2001. Asteroid 4276 was named Clifford in his honour by the Harvard-Smithsonian Center for Astrophysics in 1990.
clifford.cunningham@my.jcu.edu.au

Dorian Knight has a broad range of academic interests, with a background in both Museum Studies and Medieval Celtic/Scandinavian Studies from universities both in Britain and the Nordic Region. With a long-standing interest also in astronomy, he will shortly be starting a PhD based at the University of Iceland in Reykjavik. His doctorate will be an exploration of Old Icelandic mythic narratives in the light of myth encoding astronomical information through allegory.
drh7@hi.is

Kirk Little has been an astrologer for more than thirty years and a licensed psychiatric social worker for the past twenty-five years. Currently, he is the Clinical Director of a community mental health agency in western Maine. He has a longstanding interest in historical and philosophical aspects of astrology. He has published a number of book reviews on www.skyscript.co.uk and the essay 'Defining the Moment: Geoffrey Cornelius and the Development of the Divinatory Perspective', at http://www.astrozero.co.uk/articles/Defining.htm.
Kirklit@gmail.com

Giuliano Masola, after 40 years of work as an accountant and production planner and having been awarded the title of Master of Work, has restarted the studies at the University of Parma (Italy), gained a BA in Humanities with a dissertation in Medieval History, and is currently concluding his MA in Journalism and Media Studies. Together with Giancarlo Panicieri, he has authored the volume *Il Canale del Vescovo, secolare linfa della Val Baganza* (2013), dealing with issues of local history. He is deeply interested in astronomy and astronautics, and collaborates with some historical and literary journals.
Giuliano.masola@gmail.com

Dr. Reinhard Mussik studied Marxist-Leninist Philosophy in Leipzig, Moscow and Havana and taught Theory of Revolution at Leipzig University. After the Berlin wall fell, he continued his studies in Educational Science, Learning Psychology and Sociology. He holds a degree in Pedagogy and Philosophy from the University of Leipzig and a PhD in educational science from Humboldt University of Berlin. He's currently pursuing the MA in Cultural Astronomy and Astrology at the University of Wales, Trinity Saint David.
mussik@web.de

Günter Oestmann (b. 1959) was trained as a clockmaker and studied history of art and history of science in Tübingen und Hamburg. In 1992 he received a PhD with a study on the astronomical and astrological significance of the clock in Strasbourg Cathedral (awarded with the Philipp-Matthäus-Hahn-Prize in 1993) and in 2001 he completed a postdoctoral thesis (Habilitation) on Heinrich Rantzau and his attitude towards astrology. After having attended successfully a master class at the Federal Technical College for Clock- and Watchmakers in Karlstein/Austria he received a master craftsman's diploma in 2002. Since 2009 he is private lecturer at the Technical University Berlin. In 2013 the Musée international d'horlogerie (La Chaux-de-Fonds) awarded the "Prix Gaïa" (category history and research) to him. Fields of research: Maritime history, history of scientific instruments and clocks, history of astronomy and mathematical geography.
oestmann@nord-com.net

Nicola Reggiani holds a PhD in Greek History from the University of Parma (Italy), where he is research assistant in the same field and involved in the projects of the chair of Papyrology. He is currently also Postdoctoral *Mitarbeiter* at the Institute of Papyrology of the University of Heidelberg, and has published articles dealing with ancient Greek history (in particular, archaic Athens and the history of philosophical-political thought), Papyrology (with special regards to evidence of writing systems and writing materials in the ancient world, units of measurement in scientific and everyday use, (multi)linguistic issues), and Digital Humanities (http://unipr.academia.edu/NicolaReggiani).
nicola.reggiani@nemo.unipr.it

Karen Smyth is a Lecturer in Medieval and Early Modern Literature at the University of East Anglia. She has particular interest in the study of time markings in Middle English works (including astronomical and astrological mappings) and in East Anglian textual cultures. In her monograph *Imaginings of Time in Lydgate and Hoccleve's Verse* (Ashgate, 2011), in a co-edited interdisciplinary volume *Medieval Life-cycles: Continuity and Change* (Brepols, 2013), and in numerous articles covering works from Bede to Chaucer, she examines time as a cultural construct inherently linked to issues of authorship, authority and artistic eloquence. k.smyth@uea.ac.uk

BACK ISSUES OF CULTURE AND COSMOS
Available from *Culture and Cosmos*, PO Box 1071, Bristol BS99 1HE, U.K.
E mail culture@caol.demon.co.uk for availability and prices.

Contents, Vol. 1 no 1 (spring/summer 1997)
Robin Heath: *An Astronomical Basis for Solar Hero Myths;* **Norris Hetherington**: *Ancient Greek Cosmology and Culture: a Historiographical Review;* **Alan Weber**: *The Development of Celestial Journey Literature, 1400 - 1650;* **Ken Negus**: *Kepler's Tertius Interveniens;* **John Durant** and **Martin Bauer**: *British Public Perceptions of Astrology: an Approach from the Sociology of Knowledge.*

Contents Vol. 1 no 2 (autumn/winter 1997)
Otto Neugebauer: *On the History of Wretched Subjects;* **Nick Kollerstrom**: *The Star Zodiac of Antiquity;* **Robert Zoller**: *The Hermetica as Ancient Science;* **Edgar Laird**: *Christine de Pizan and Controversy Concerning Star Study in the Court of Charles V;* **Jürgen G.H. Hoppman**: *The Lichtenberger Prophecy and Melanchthon's Horoscope for Luther;* **Elizabeth Heine**: *W.B.Yeats: Poet and Astrologer.*

Contents Vol. 2 no 1 (spring/summer 1998)
J. McKim Malville and **R. N. Swaminathan**: *People, Planets and the Sun: Surya Puja in Tamil Nadu, South India;* **Carlos Trenary**: *Yaxchilan Lintel 25 as a Cometary Record;* **Graziella Federici Vescovini**: *Biagio Pelacani's Astrological History for the Year 1405;* **Frank McGillion**: *The Influence of Wilhelm Fliess' Cosmobiology on Sigmund Freud;* **Nicholas Campion**: *Sigmund Freud's Investigation of Astrology.*

Contents Vol. 2 no 2 (autumn/winter 1998)
Giuseppe Bezza: *Astrological Considerations on the Length of Life in Hellenistic, Persian and Arabic Astrology;* **Angela Voss**: *The Music of the Spheres: Marsilio Ficino and Renaissance harmonia*; **Robert Zoller**: *Marc Edmund Jones and New Age Astrology in America.*

Contents Vol. 3 no 1 (spring/summer 1999)
Michael R. Molnar: *Firmicus Maternus and the Star of Bethlehem*; **Roger Beck**: *The Astronomical Design of Karakush, a Royal Burial Site in Ancient Commagene: an Hypothesis*; **Chantal Allison**: *The Ifriqiya Uprising Horoscope from* On Reception *by Masha'alla, Court Astrologer in the Early 'Abassid Caliphate.*

Contents Vol. 3 no 2 (autumn/winter 1999)
Robin Waterfield: *The Evidence of Astrology in Classical Greece;* **Remo Catani**: *The Polemics on Astrology 1489-1524*; **Claudia Rousseau**: *An Astrological Prognostication to Duke Cosimo de Medici of Florence.*

Contents Vol. 4 no 1 (spring/summer 2000)
Patrick Curry: *Historical Approaches to Astrology*; **Edgar Laird**: *Heaven and the Sphaera Mundi in the Middle Ages*; **George D. Chryssides**: *Is God a Space Alien? The Cosmology of the Raëlian Church.*

Contents Vol. 4 no 2 (autumn/winter 2000)
David J. Ross: *The Bird, The Cross, and the Emperor: Investigations into the Antiquity of The Cross in Cygnus*; **Angela Voss:** *The Astrology of Marsilio Ficino: Divination or Science?*; **Patrick Curry:** *Astrology on Trial, and its Historians: Reflections on the Historiography of 'Superstition'*.

Contents Vol. 5 no 1 (spring/summer 2001)
Demetra George: *Manuel I Komnenos and Michael Glykas: A Twelfth-Century Defence and Refutation of Astrology*, Part 1; **Richard L. Poss:** *Stars and Spirituality in the Cosmology of Dante's* Commedia.

Contents Vol. 5 no 2 (autumn/winter 2001)
Arkadiusz Sołtysiak: *The Bull of Heaven in Mesopotamian Sources*; **Demetra George:** *Manuel I Komnenos and Michael Glykas: A Twelfth-Century Defence and Refutation of Astrology*, Part 2; **Garry Phillipson** and **Peter Case:** *The Hidden Lineage of Modern Management Science: Astrology, Alchemy and the Myers-Briggs Type Indicator*.

Contents Volume 6 Number 1 (spring/summer 2002)
Ari Belenkyi: *A Unique Feature of the Jewish Calendar - Deĥiyot*; **Demetra George:** *Manuel I Komnenos and Michael Glykas: A Twelfth-Century Defence and Refutation of Astrology*, Part 3; **Germana Ernst** : *The Sky in a Room: Campanella's Apologeticus in defence of the pamphlet* De siderali fato vitando; **Tommaso Campanella:** *Apologia for the opuscule on* De siderali fato vitando.

Contents Volume 6 Number 2 (autumn/winter 2002)
Jesse Krai: *Rheticus' Poem* 'Concerning the Beer of Breslau and the Twelve Signs of the Zodiac'; **Anna Marie Roos:** *Israel Hiebner's Astrological Amulets and the English Sigil War*; **Nicholas Campion:** *Surrealist Cosmology: André Breton and Astrology*.

Contents Volume 7 Number 1 (spring/summer 2003) GALILEO'S ASTROLOGY
Nick Kollerstrom: *Foreword: Galileo as Believer*; **Nicholas Campion**: *Introduction: Galileo's Life and Work*; **Antonio Favaro**: *Galileo, Astrologer*; **Germana Ernst**: *Astrology and Prophecy in Campanella and Galileo*; **Nick Kollerstrom**; *Galileo as an Astrologer*: Antonino Poppi: *On Trial for Astral Fatalism: Galileo Faces the Inquisition*; **Guiseppe Righini**:*Galileo's Horoscope for Cosimo II de Medici*; **Mario Biagioli**: *An Astrologico-Dynastic Encounter*; *Galileo's Correspondence*; *Galileo's Letter to Dini, May 1611*; *On the Character of Sagredo: Galileo's judgements upon his nativity*; *Galileo's Horoscopes for his Daughters*; *Rome, 1630*; **Bernadette Brady**: *Four Galilean Horoscopes: An Analysis of Galileo's Astrological Techniques*; *A Sonnet by Galileo*.

Contents Volume 7 Number 2 (autumn/winter 2003)
Günther Oestmann: *Tycho Brahe's Geniture*; **Bernard Eccles**: *Astrological physiognomy from Ptolemy to the present day*; **James Brockbank**: *Planetary signification from the second century until the present day*; **Julia Cleave**: *Ficino's Approach to Astrology as Reflected in Book VII of his Letters*.

Contents Volume 8 No 1/2 (spring/summer autumn/winter 2004)
Valerie Shrimplin *Organising INSAP*; **Rolf Sinclair** *Foreword: INSAP IV in Oxford: A Summary*; **Nicholas Campion** *Introduction: The Inspiration of Astronomical Phenomena*:

Hubert A. Allen, Jr. *Hawkins' Way: Remembering Astronomer Gerald S. Hawkins*; **Hubert A. Allen, Jr. and Terry Edward Ballone** *Star Imagery in Petroglyph National Monument*; **Mark Butterworth** *Astronomy and the Magic Lantern*; **Ann Laurence Caudano** *Sun, Moon, and Stars on Kievan Rus Jewellery ($10^{th} - 13^{th}$ Centuries)*; **Nicholas Campion** *The Sun is God;* **Anne Chapman-Rietschi** *Cosmic Gardens*; **Deborah Garwood** *Paris Solstice*; **N. J. Girardot** *Celestial Worlds In the Work of Self-Taught Visionary Artists With Special Reference to Howard Finster's Vision of 1982*; **John G. Hatch** *Desire, Heavenly Bodies, and a Surrealist's Fascination with the Celestial Theatre*; **Holly Henry** *Bertrand Russell in Blue Spectacles: His Fascination with Astronomy*; Ronald Hicks *Astronomy and the Sacred Landscape in Irish Myth*; **Chris Impey** *Why Are We So Lonely?*; **Bernd Klähn** *The Aberration of Starlight and/in Postmodernist Fiction*; **Nick Kollerstrom** *How Galileo dedicated the moons of Jupiter to Cosimo II de Medici*; **Arnold Lebeuf** *Dating the five Suns of Aztec cosmology*; **Andrea D. Lobel** *Trailing the Paper Moon: Astronomical Interpretations of Exodus 12:1-2*; **Stephen C. McCluskey** *Wordsworth's 'Rydal Chapel' and the Astronomical Orientation of Churches*; **David Madacsi** *Sky: Atmospheres and Aesthetic Distance in Planetary and Lunar Environments*; **Daniel R. Matlaga** *A Journey of Celestial Lights: The Sky as Allegory in Melville's Moby Dick*; **Paul Murdin** *Representing the Moon*; **R. P. Olowin** *Robinson Jeffers: Poetic Responses to a Cosmological Revolution*; **David W. Pankenier** *A Brief History of Beiji (Northern Culmen)*; **Richard Poss** *Poetic Responses to the Size of the Universe: Astronomical Imagery and Cosmological Constraints*; **Barbara Rappenglück** *The material of the solid sky and its traces in cultures*; **Brad Ricca** *The Night of Falling Stars: Reading the 1833 Leonid Meteor Storm*; **Patricia Ricci** *Lux ex Tenebris: Etienne-Louis Boullée's Cenotaph for Sir Isaac Newton*; **Sarah Richards** *Die Planetentheorie: its uses and meanings for the Saxon mining communities and the culture of the Dresden Court 1553-1719*; **William Saslaw and Paul Murdin** *The Double Apollos of Istrus*; **Petra G. Schmidl** *Dusk and Dawn in Medieval Islam; On the Importance of Twilight Phenomena with Some Examples of Their Representations in Texts and on Instruments*; **Valerie Shrimplin** *Borromini and the New Astronomy: the elliptical dome*; **Joshua Stein** *Cicero's Use of Astronomy as Proof of the Existence of the Gods*; **Antje Steinhoefel** *Art and Astronomy in the Service of Religion:Observations on the Work of John Russell (1745-1806)*; **Burkard Steinrücken** *An interpretation of the 'Sky Disc of Nebra' as an icon for a bronze age planetarium mechanism with parallels to the moving world-soul in Plato's* Timaeus; **Gary Wells** *Daumier and The Popular Image of Astronomy.*

Contents Vol. 9 no 1 (Spring/Summer 2005)
Gennadij Kurtik and Alexander Militarev *Once more on the origin of Semitic and Greek star names:an astronomic-etymological approach updated*; **Prudence Jones** *A Goddess Arrives: Nineteenth Century Sources of the New Age Triple Moon Goddess*; **Louise Curth** *Astrological Medicine and the Popular Press in Early Modern England.*

Contents Vol. 9 no 2 (Autumn/Winter 2005)
Marinus Anthony van der Sluijs *A Possible Babylonian Precursor to the Theory of ecpyrōsis*; **Liz Greene** *Did Orphic Beliefs Influence the Development of Hellenistic Astrology?*; **Ariel Cohen** *Astronomical Luni-Solar Cycles and the Chronology of the Masoretic Bible*; **Tayra Lanuza-Navarro** *An Astrological Disc from the Sixteenth Century*; **J.C. Holbrook** *Celestial Navigators and Navigation Stories.*

Contents Vol. 10 no 1 and 2 (Spring/Summer, Autumn/Winter 2006)

Lucia Dolce *Introduction: The worship of celestial bodies in Japan: politics, rituals and icons*; **Lucia Dolce** *The State of the Field: A basic bibliography on astrological cultic practices in Japan*; **Hayashi Makoto** *The Tokugawa Shoguns and Yin-yang knowledge (onmyōdō)*; **John Breen** *Inside Tokugawa religion: stars, planets and the calendar-as-method*; **Mark Teeuwen** *The imperial shrines of Ise:An ancient star cult?*; **Lilla Russell-Smith** *Stars and Planets in Chinese and Central Asian Buddhist Art from the Ninth to the Fifteenth Centuries*; **Matsumoto Ikuyo** *Two Mediaeval Manuscripts on the Worship of the Stars from the Fujii Eikan Collection*; **Tsuda Tetsuei** *The Images of Stars and Their Significance in Japanese Esoteric Buddhist Art*; **Meri Arichi** *Seven Stars of Heaven and Seven Shrines on Earth: The Big Dipper and the Hie Shrine in the Medieval Period*; **Gaynor Sekimori** *Star Rituals and Nikko Shugendō*; **Meri Arichi** *The front cover image: Myōken Bosatsu.*

Contents Vol. 11 no 1 and 2 (Spring/Summer, Autumn/Winter 2007)
Micah Ross *A Survey of Demotic Astrological Texts*; **Francis Schmidt** *Horoscope, Predestination and Merit in Ancient Judaism*; **Stephan Heilen** *Ancient Scholars on the Horoscope of Rome*; **Joanna Komorowska** *Philosophy among Astrologers*; **Wolfgang Hübner** *The Tropical Points of the Zodiacal Year and the Paranatellonta in Manilius' Astronomica*; Aurelio Pérez Jiménez *Hephaestio and the Consecration of Statues*; **Robert Hand** *Signs as Houses (Places) in Ancient Astrology*; **Dorian Gieseler Greenbaum** *Calculating the Lots of Fortune and Daemon in Hellenistic Astrology*; **Susanne Denningmann** *The Ambiguous Terms ⌧⌧α and ⌧σπερία, ⌧νατολή, and ⌧⌧α and ⌧σπερία δύσις* **Joseph Crane** *Ptolemy's Digression: Astrology's Aspects andMusical Intervals*; **Giuseppe Bezza** *The Development of an Astrological Term – from Greek* hairesis *to Arabic* ⌧ayyiz; **Deborah Houlding** *The Transmission of Ptolemy's Terms: An Historical Overview, Comparison and Interpretation.*

Contents Vol. 12 no 1 (Spring/Summer 2008)
Liz Greene *Is Astrology a Divinatory System?*; **James Maffie** *Watching the Heavens with a 'Rooted Heart': The Mystical Basis of Aztec Astronomy*; **J.C. Holbrook** *Astronomy and World Heritage.*

Contents Vol. 12 no 2 (Autumn/Winter 2008)
Mark Williams *Astrological Poetry in late medieval Wales: the case of Dafydd Nanmor's 'To God and the planet Saturn'*; **Scott Hendrix** *Choosing to be Human: Albert the Great on Self Awareness and Celestial Influence*; **Graham Douglas** *Luis Vilhena and the World of Astrology.*

Contents Vol. 13 no 1 (Spring/Summer 2009)
Josefina Rodríguez-Arribas *Astronomical and Astrological Terms in Ibn Ezra's Biblical Commentaries: A New Approach*; **Andrew Vladimirou** *Michael Psellos and Byzantine Astrology in the Eleventh Century*; **Marinus Anthony van der Sluijs** *The Dragon of the Eclipses—A Note*; **Patrick Curry** *Response to Liz Greene's 'Is Astrology a Divinatory System?'*

Contents Vol. 13 no 2 (Autumn/Winter 2009)
Liz Greene *Mystical Experiences Among Astrologers*; **Peter Pesic** *How the Sun Stood Still: Old English Interpretations of Joshua and the Leap Year*; **Doina Ionescu** *Virginia Woolf and Astronomy*; **Carlos Ziller Camenietzki and Luis Miguel Carolino** *Astrologers at*

War: Manuel Galhano Lourosa and the Political Restoration of Portugal, 1640–1668; **Nick Campion** *Astrology's Role in New Age Culture: A Research Note*

Contents Vol. 14 no 1 and 2 (Spring/Summer, Autumn/Winter 2010)
Dorian Gieseler Greenbaum *Introduction*; **Friederike Boockmann** *Johann Kepler's Horoscope Collection*; **J. Cornelia Linde (trans.)** *Helisaeus Röslin's Delineation of Kepler's Birthchart, 1592*; **J. Cornelia Linde and Dorian Greenbaum (trans.)** *David Fabricius and Kepler on Kepler's Personal Astrology, 1602*; **Dorian Greenbaum (trans.)** *Kepler's Delineation of his Family's Astrology*; **J. Cornelia Linde and Dorian Greenbaum (trans.)** *Kepler and Michael Mästlin on their Son's Nativities, 1598*; **J. Cornelia Linde and Dorian Greenbaum (trans.)** *Kepler's Methods of Astrological Interpretation for Rudolf II, 1602*; **J. Cornelia Linde and Dorian Greenbaum (trans.)** *Kepler's Astrological Interpretation of Rudolf II by Traditional Methods, 1602*; **J. Cornelia Linde and Dorian Greenbaum (trans.)** *Kepler's Letter to an Official on Rudolf II and Astrology, 1611*; **J. Cornelia Linde and Dorian Greenbaum (trans.)** *Excerpts from Kepler's Correspondence and Interpretation of Wallenstein's Nativity, 1624-1625*; **J. Cornelia Linde and Dorian Greenbaum (trans.)** *The Nativities of Mohammed and Martin Luther, 1604*; **J. Cornelia Linde and Dorian Greenbaum (trans.)** *The Nativity of Augustus*; **John Meeks** *Introduction: Kepler and the Art of Weather Prognostication*; **John Meeks (trans.)** *Kepler's Weather Calendar of 1618*; **John Meeks (trans.)** *Excerpts from Kepler's Weather Calendar of 1619*; **Patrick J. Boner (trans.)** *Astrology on Trial: Kepler, Pico and the Preservation of the Aspects De stella nova: Chapters 7-9*; **J. Cornelia Linde and Dorian Greenbaum (trans.)** *On Directions*; **J. Cornelia Linde and Dorian Greenbaum (trans.)** *David Fabricius and Kepler on Astrological Theory and Doctrine, 1602*; **J. Cornelia Linde and Dorian Greenbaum (trans.)** *David Fabricius and Kepler on Fabricius's Directions, 1603-1604*; **J. Cornelia Linde and Dorian Greenbaum (trans.)** *On Aspects, 1602*; **Appendix** *A Selection of Kepler's Handwritten Charts*

Contents Vol. 15 no 1 (Spring/Summer 2011)
Miguel Querejeta *On the Eclipse of Thales, Cycles and Probabilities*; **Nicholas Campion** *The Shock of the New: The Age of Aquarius*; **Alejandro Gangui** *The Barolo Palace: Medieval Astronomy in the Streets of Buenos Aires*; **Nicholas Campion and John Frawley** *Research Note: A Horoscope by André Breton*

Contents Vol. 15 no 2 (Autumn/Winter 2011)
Liz Greene *Heavenly Hosts: Angelic Intermediaries as Soul-Gates*; **Pamela Armstrong** *Ritual Ornamentation—From the Secular to the Religious*; **Paul Cheshire** *William Gilbert: Macrocosmal Astrologer in an Age of Revolution*; **Sylwia Konarska-Zimnicka** *Astrologia Licita? Astrologia Illicita? The Perception of Astrology at Kraków University in the Fifteenth Century*; **John Frawley** *Research Note: William Blake and Antares*

Contents Volume 16 No 1/2 (spring/summer autumn/winter 2012)
Nicholas Campion, *Editorial: The Inspiration of Astronomical Phenomena*; **Chris Impey**, *The Inspiration of Astronomical Phenomena*; **Ulisses Barres de Almeida**, *What are these sparks of infinite clarity? And what am I? So I pry*; BATH AND THE HERSCHELS: **Michael Hoskin**, *William Herschel's Wonderful Decade, 1781–1790*; **Francis Ring**, *The Bath Philosophical Society and its influence on William Herschel's career*; **Roberta J.M. Olson and Jay M. Pasachoff**, *The Comets of Caroline Herschel, Sleuth of the Skies at Slough*; HISTORY AND CULTURE: **V.F. Polcaro and A. Martocchia**, *Guidelines for a social history*

of Astronomy; **Euan MacKie**, *A new look at the astronomy and geometry of Stonehenge*; **Leonid Marsadolov**, *Archaeoastronomical Aspects of the Archaeological Monuments of Siberia*; **Christian Etheridge**, *A systematic re-evaluation of the sources of Old Norse astronomy*; **Aidan Foster**, *Hierophanies in the Vinland Sagas: Images of a New World*; **Inga Elmqvist Söderlund**, *Inspiration from antique heroic deeds: Hercules as an astronomer*; **Patricia Aakhus**, *Astral Magic and Adelard of Bath's Liber Prestigiorum; or Why Werewolves Change at the Full Moon*; **David Pankenier**, Astrology for an Empire: The 'Treatise on the Celestial Offices' (ca. 100 BCE); **Steven Renshaw**, *The Inspiration of Subaru as a Symbol of Values and Traditions in Japan*;b **Daniel Armstrong**, *Citing The Saucers: Astronomy, UFOs and a persistence of vision*; **Alberto Cappi**, *The concept of gravity before Newton*; **Paul Murdin**, *Artilleryman to head of state—how astronomy inspired Francois Arago*; **Paolo Molaro and Alberto Cappi**, *Edgar Allan Poe's cosmology in* Eureka; **Voula Saridakis**, *For 'the present and future happiness of my dear Pupils'": The Astronomical and Educational Legacy of Margaret Bryan*; **Michael Rowan–Robinson**, *The invisible universe*; THE ARTS: **Arnold Wolfendale**, *The Inter-Relation of the Visual Arts and Science in General and Astronomy in Particular*; **Lynda Harris**, *Changing Images of the Milky Way during the Greco-Roman and Medieval Periods*; **Lucia Ayala**, *The Universe in images: Iconography of the Plurality of Worlds*; **Tayra M. Carmen Lanuza-Navarro**, *Astrological culture before its public: the representation of astrology in Golden Age Spanish Theatre*; **Emily Urban**, *Depicting the Heavens: The Use of Astrology in the Frescoes of Rome*; **Michael Mendillo**, *The Artistic Portrayal of the Medicean Moons in Early Astronomical Charts, Books and Paintings*; **Rolf Sinclair**, *Howard Russell Butler: Painter Extraordinary of Solar Eclipses*; **Beatriz Garcia, Estela Reynoso, Silvina Pérez Alvarez and Rubén Gabellone**, *Inspiration of Astronomy in the movies: a history of a close encounter*; **Gary Wells**, *The Moon in the Landscape: Interpreting a Theme of 19th Century Art*; **Clive Davenhall**, *The Space Art of Scriven Bolton*; **Matthew Whitehouse**, *Astronomical Organ Music*; **Aaron Plasek**, *Between Scientists, Writers and Artists: Theorising and Critiquing Knowledge-Production at the Interstices between Disciplines*; ARTISTS: **Merja Markkula**, *The Way I See the Stars: fibre art inspired by astrobiology*; **Govinda Sah**, *Beyond the Notion*; **Gisela Weimann**, *Above all the stars*; **Courtney Wrenn**, *Nebulae (emission / absorption)*; **Toby MacLennan**, *Presentation of Playing the Stars*; **Felicity Spear**, *Extending vision: sky-situated knowledge and the artist's eye*; **Vanessa Stanley**, *Surveillance-Surveillance-Surveillance*; **Jim Cogswell**, *Molecular Delirium*.

www.ingramcontent.com/pod-product-compliance
Lightning Source LLC
Chambersburg PA
CBHW070953080526
44587CB00015B/2288